Louis 6.

No. 912
$7.95

88 PRACTICAL OP AMP CIRCUITS YOU CAN BUILD

BY GEORGE B. CLAYTON
DEPARTMENT OF PHYSICS
LIVERPOOL POLYTECHNIC

FIRST EDITION

FIRST PRINTING—APRIL 1977

Originally published in 1975 by
THE MACMILLAN PRESS LTD., London and Basingstoke

Library of Congress Cataloging in Publication Data

Clayton, George Burbridge
    88 Practical OP AMP circuits you can build.

    British ed. published in 1976
    Under title: Experiments with operational amplifiers.
    Bibliography: p.
    Includes index.
    1. Operational amplifiers. 2. Electronic circuits. 1. Title.
TK7871.58.06C52   1977   621.3815'35   77-7500

ISBN 0-8306-7912-X
ISBN 0-8306-6912-4 pbk.

Preface

Integrated circuit operational amplifiers are now so cheap that it is economically possible to use them freely in all types of electronic instrumentation. One can afford to think of them as some kind of 'super transistor' but their use considerably simplifies the design and construction of circuits compared with the use of discrete components and transistors. Also, operational amplifier modules make it easier for the non-electronics specialist to construct the instrumentation circuits needed for his particular field of work.

This book covers a range of practical operational amplifier applications, gives circuits which include component values and suggests measurements that can be made in order to study circuit action. The book is intended as an experimental supplement to the author's previous book on operational amplifiers[1].

It is suggested that the quickest way for the non-electronics specialist to learn about operational amplifiers is actually to use them in working circuits. It does not matter very much if a wrong connection is made in the experimental circuits, the operational amplifier type suggested for use will tolerate quite a few mistakes and even if you destroy it it should not break you. If resistor values suggested in the circuits are not at hand try other values; electronic systems will work (in a fashion) with a considerable range of component values. Having investigated the practical circuits, then is the time to start reading to find out in more detail how the circuits work and how their performance can be refined. This alternation between experimental work and reading is most rewarding. Experimental work increases confidence and leads to sounder understanding, the reading suggests further experimental work which when carried out encourages further theoretical exploration.

The experiments suggested in the book cover a wide range of operational amplifier applications which will be found useful for a variety of measurement and instrumentation systems. The emphasis is on working circuits designed to give a practical understanding of the principles underlying each application. The way in which performance errors are related to the characteristics of the particular amplifier used in the circuit are briefly treated in an Appendix.

This book should prove useful for all those interested in experimentally finding out about the capabilities of operational amplifiers. A prerequisite for the reading of the book extends little beyond a knowledge of basic d.c. and a.c. circuit theory.

References

1. G.B. Clayton. *Operational Amplifiers,* Butterworths (1971)

Contents

1. Basic Operational Amplifier Ideas

In this chapter a brief summary of elementary operational amplifier concepts is given, prior to the main body of the text which is concerned with experimental investigations of operational amplifier applications.

1.1 Introduction

A differential input operational amplifier is a device with two input terminals, an output terminal and two power supply terminals. In addition, it will normally have terminals for offset balancing (for setting the output to zero when the input is zero) and may have terminals to which external components can be connected in order to modify the frequency response characteristics of the amplifier.

The circuit symbol for an operational amplifier is a horizontal triangle; operational amplifiers are used with dual power supplies consisting of a positive and negative supply in series connected as shown in figure 1.1

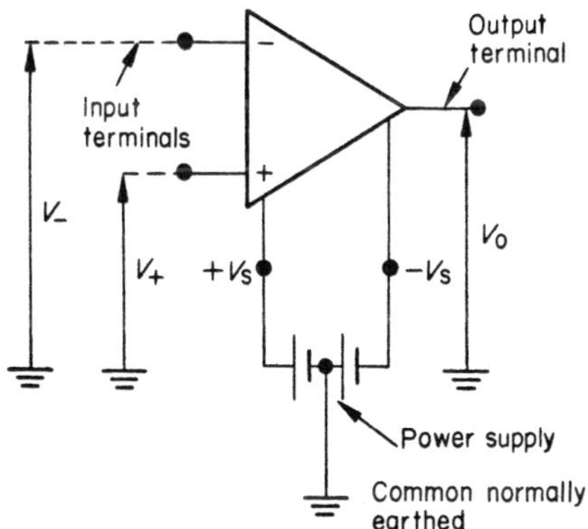

Fig. 1.1   Circuit symbol used for an operational amplifier

The two input terminals of an operational amplifier are normally distinguished from each other symbolically by a (+) and (−) sign. Input and output signals are measured with respect to the power supply common terminal which is normally earthed. Note that the (+) and (−) notation used at the input *does not* mean positive voltages go into one terminal and negative at the other, it means that signals applied to the (−) terminal cause signal changes of opposite polarity at the output, signals applied to the (+) terminal cause signal changes of the same polarity at the output. The (−) terminal inverts, the (+) terminal is non-inverting.

1.2 Open Loop Gain

The output signal of an operational amplifier is controlled by the difference in the signals applied to the (+) and (−) input terminals. We may write

$$V_0 = A_{OL}\,(V_+ - V_-) \tag{1.1}$$

A_{OL} is called the open loop gain, it is very large, even the most modest of operational amplifiers having open loop gains which exceed 10^4 (80 dB).

1.3 Open Loop Transfer Curve − Output Voltage Limits

The relationship between the output voltage and input difference voltage can be shown graphically in the form of an open loop transfer curve. A slightly idealised transfer curve for an operational amplifier is shown in figure 1.2. Note that only a very small change in the input difference voltage is required to cause the output to go between its saturated levels.

The maximum output voltage that an operational amplifier can give is usually limited to a volt or so less than the applied power supply voltages. The slope of the transfer curve gives a value for the open loop gain of the amplifier.

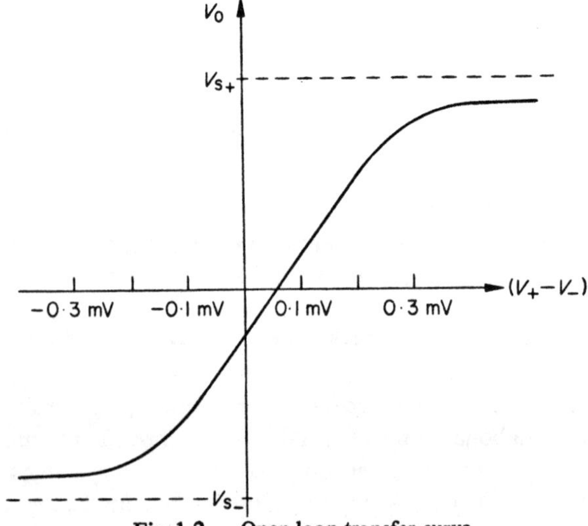

Fig. 1.2　　Open loop transfer curve

Practical operational amplifiers have transfer curves which are often considerably less linear than that shown in figure 1.2. Also, in the case of a practical amplifier the curve does not usually pass through the origin. A small input difference voltage has to be applied to a practical amplifier to make its output zero;this is called the input offset voltage. Input offset voltages are typically 1 millivolt. This may appear large in comparison with the input difference voltage required to cause the output to go between maximum limits, but in the practical negative feedback circuits in which operational amplifiers are used amplifier non-linearities and offsets do not have a first order effect.

1.4 Properties of an Almost Ideal Operational Amplifier

In addition to large open loop gain, operational amplifiers are normally designed to have very high input impedance and low output impedance. Other performance parameters of operational amplifiers such as bias current, common mode rejection ratio, frequency response characteristics and slewing rate are also important but for practical purposes we confine our attention to a somewhat idealised amplifier.

In this simplified first treatment an operational amplifier is assumed to have the following properties

(1)   A very large open loop gain.
(2)   Output unaffected by signal frequency, no signal phase shift with change in frequency.
(3)   A very large input impedance so that the amplifiers takes negligible currents at its input terminals.
(4)   A very small output impedance so that the output of the amplifier is unaffected by loading.
(5)   Zero output voltage for zero input voltage (offset zero).

1.5 Usefulness of an Operational Amplifier. What Can It Do?

The usefulness of an operational amplifier stems from the constraints which it can impose on passive networks externally connected to it. These constraints arise as a direct result of the foregoing properties. In operational amplifier applications a negative feedback path is normally connected in some way between the output terminal of the amplifier and the inverting input terminal. Connected in this way the output voltage of the amplifier always changes in such a way as to reduce the input difference voltage to a very small value. Some of the various circuit configurations are now discussed. For convenience power supplies will not be shown in these first circuits.

1.5.1 *Unity Gain Follower*

In the circuit shown in figure 1.3 the output terminal of the amplifier is connected directly to the inverting input terminal making $V_- = V_0$ thus using equation 1.1.

$$V_0 = A_{OL} (V_{in} - V_0)$$

Therefore

$$V_{in} - V_0 = \frac{V_0}{A_{OL}} \rightarrow 0 \text{ (Since } A_{OL} \text{ is very large)}$$

Thus

$$V_0 = V_{in}$$

Note that the output signal returned to the inverting input terminal takes on that value which is required to force the input difference voltage towards zero.

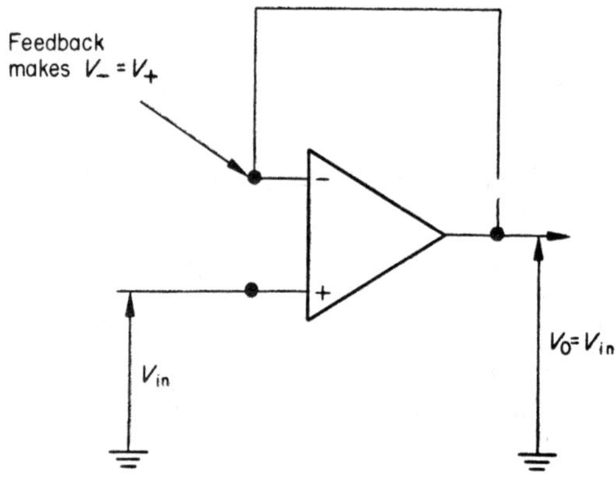

Fig. 1.3   Unity gain follower

In figure 1.3 the output voltage will always be within V_0/A_{OL} of the input voltage V_{in}. The error by which V_0 departs from V_{in} is greatest when V_0 has its maximum value of, say, 14 volts but the error is then only $14/10^5 = 0.14$ millivolts if the open loop gain of the amplifier is 10^5. The importance of the unity gain follower circuit of figure 1.3 arises from its impedance characteristics. It has a very high input impedance and a very low output impedance and serves as an excellent 'buffer' stage preventing interaction between a signal source and load.

1.5.2 *The Follower With Gain*

In the circuit of figure 1.4 the output signal is attenuated by a resistive divider (by the ratio $R_1/(R_1 + R_2)$) before being applied to the inverting input terminal. Remember it is assumed that no current flows into either input terminal of the amplifier. The output voltage, as before, takes on that value required to force the input difference voltage to zero it must thus have a value which is the inverse of the attenuation ratio multiplied by the input signal V_{in}. *Note that an infinite open loop gain makes the gain of the feedback operational amplifier circuit entirely dependent upon the external circuit elements.*

1.5.3 *The Inverter*

If in the circuit of figure 1.4 we earth the (+) input terminal of the amplifier and apply the input signal to the end of the resistor R_1 ; the circuit inverts the polarity of the applied input signal but the gain of the feedback circuit is still determined only by the value of the external circuit elements. In figure 1.5 the output voltage forces the voltage at the (−) input towards zero and the following relationships are established.

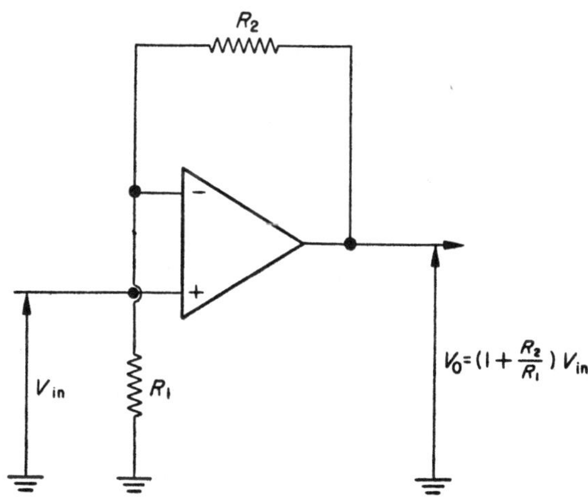

Fig. 1.4   Follower with gain

The current supplied by the input signal is

$$I_{in} = \frac{V_{in}}{R_1}$$

The effective input resistance of the circuit is thus R_1. The output voltage causes I_{in} to flow through resistor R_2 and

$$\frac{V_{in}}{R_1} = I_{in} = I_f = -\frac{V_o}{R_2}$$

Thus

$$V_o = -\frac{R_2}{R_1} V_{in}$$

1.5.4 *Summing Amplifier*

In figure 1.5 the output voltage causes any current arriving at the minus input to flow through the feedback path. If a number of input voltages are connected in series with the resistors that meet at the (−) input terminal, the sum of the currents passing through these resistors will be made to flow through the feedback resistor. The arrangement is shown in figure 1.6. We have

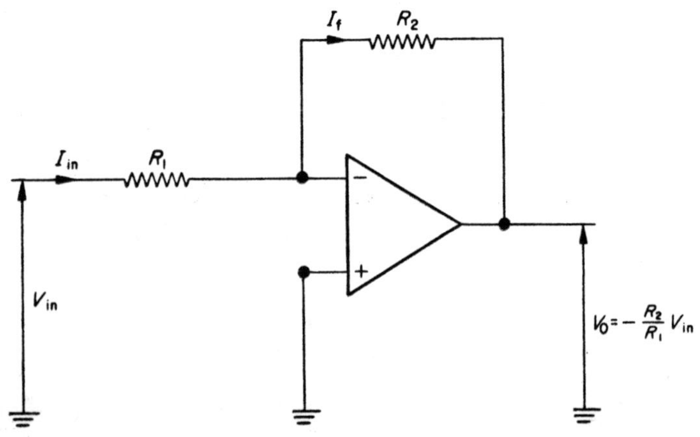

Fig. 1.5   Inverting amplifier

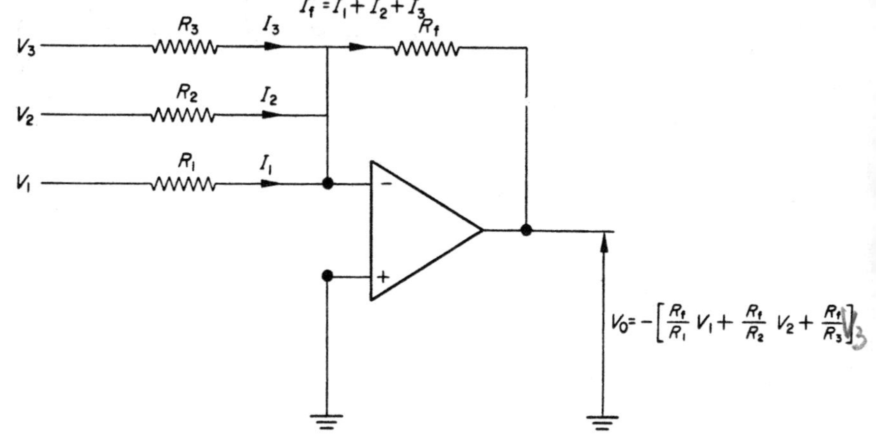

Fig. 1.6   Summing amplifier

$$I_1 + I_2 + I_3 = I_f$$

Thus

$$\frac{V_1}{R_1} + \frac{V_2}{R_2} + \frac{V_3}{R_3} = -\frac{V_0}{R_f}$$

or

$$V_0 = -\left[\frac{R_f}{R_1} V_1 + \frac{R_f}{R_2} V_2 + \frac{R_f}{R_3} V_3\right]$$

An operational amplifier can thus be used to sum a number of voltages or currents independently; the (−) terminal of the amplifier is often called the amplifier summing point.

1.5.5 *Integrator*

Thus far only resistive circuits have been considered but many other kinds of element can be externally connected to an operational amplifier in order to obtain a required input output relationship. In figure 1.7 the feedback resistor has been replaced by a capacitor. As before, the output voltage takes on that value required to cause the current arriving at the summing point to flow through the feedback path, it must do this in order to force the input difference voltage to zero and the current has nowhere else to go.

The input current $I_{in} = \dfrac{V_{in}}{R_1}$ flows into the feedback capacitor and charges it up. Thus

$$I_{in} = \frac{dq}{dt} = C \frac{dV_c}{dt} = -C \frac{dV_0}{dt}$$

and

$$\frac{dV_0}{dt} = -\frac{1}{CR_1} V_{in}$$

or

$$V_0 = -\frac{1}{CR_1} \int V_{in} \; dt$$

The output of the amplifier is proportional to the integral with respect to time of the input signal.

1.5.6 *Further Uses*

The foregoing sections cover only a few of the operational amplifier applications that have been devised and which remain to be discovered by the ingenious operational amplifier user. Interchanging the resistance and capacitance in figure 1.7 gives a differentiator circuit, the use of non linear input or feedback elements

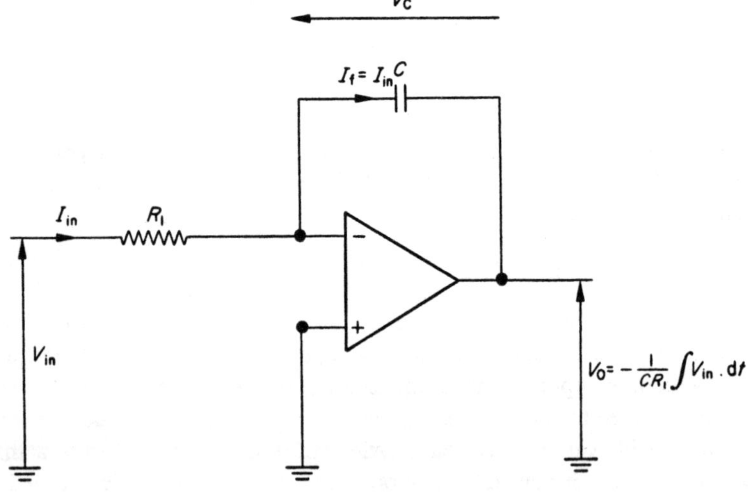

Fig. 1.7   Integrator circuit

gives non linear input output relationships. Both input terminals of a differential input operational amplifier may have signals applied to them so as to give a subtractor operation. Operational amplifiers may be used with positive feedback to give a multivibrator action, as comparators, and to generate a variety of signal waveforms including triangular waves, square waves and sinusoidal waves.

Practical circuits illustrating many of the uses of operational amplifiers are given in the following chapters of the book. The best way to learn about operational amplifiers is to use them, we suggest that the reader should now connect up the circuits given in Chapter 2. Real amplifiers do not behave exactly as the ideal amplifier considered in the previous sections but differences are small and departures between real and ideal are conveniently treated as performance errors. Do not worry initially about these performance errors, test the response of the circuits to a variety of signals, try the effect of changing component values, and generally gain familiarity with the circuits. Having done this if you need to know more about the circuits, or if you want to refine their performance by using a different amplifier type you should return to a more detailed study of real amplifier characteristics[1].

References

1. G.B. Clayton, *Operational Amplifiers.* Butterworths (1971).

2. Basic Operational Amplifier Applications

2.1 The Amplifier Used Experimentally

This chapter deals with basic practical applications of operational amplifiers. There are many types available and the different amplifiers are normally distinguished by having specific performance characteristics which are optimised for use in particular kinds of application. In order to simplify experimental procedure the circuits given in the book all use the same amplifier type, the 741. This type was chosen for its ready availability, for the circuit protection which it incorporates and for the fact that it requires no external frequency compensation in most applications. The internal circuit protection means that amplifier damage caused by an inadvertent wrong connection is likely to be avoided. Internal frequency compensation, although it restricts the amplifier slewing rate, simplifies experimental circuits and makes closed loop circuits less prone to instability. The 'slowing down' of the amplifier is no real disadvantage for experimental purposes, it serves to emphasise those applications in which amplifier slewing rate is a limiting parameter. If the experimental circuits suggested in the book are to be developed as the basis of practical applications it may be necessary to use an alternative amplifier type in order to meet the requirements of some limiting performance specification. The reader is referred to the appendix and to the author's earlier book, 'Operational Amplifiers', for a treatment of the considerations involved in the selection of an amplifier type for a particular application.

The 741 device is available both in metal can and dual in line plastic packages. There are a large number of suppliers of this device and for the purpose of the experimental evaluations of applications suggested in the book it does not really matter which manufacturers' 741 is used. However, it is worth noting that minor processing differences between manufacturers can, and does, make 741's which are not all the same, some 741's are definitely better than others. They differ in offset and drift specifications and in a.c. performance, some are said to be faster than a 'normal' 741. All this means is that the internal frequency compensating capacitor, which is formed on the device chip, has a smaller value that normal. Whilst this does give a faster slewing rate it means that there is less phase margin against instability in closed loop circuit configurations. Amplifiers also differ in noise performance, in particular with respect to the amount of so-called 'popcorn noise' which they give. Popcorn noise is the erratic jump of bias current between two levels which will be found to take place at random intervals in some amplifiers. The differences mean that it is perhaps a mistake to choose simply the cheapest 741 for use in an application which is in any way critical; by paying more one can get a device with tighter specifications (for example, Analog Devices AD 741, J, K, L and S devices).

The circuits given in the book are most rapidly connected up if some form of circuit breadboard assembly is employed. The author used commercially available circuit boards, T-decs, boards specifically intended for use with multipin i.c.'s. Most of the circuits require the use of twin power supplies. When using separate positive and negative supplies, signals are measured with respect to the potential of the power supply common line which is normally grounded.

2.2 Resistive Feedback Circuits

In most of the practical applications of an operational amplifier the amplifier is connected in some form of negative feedback circuit. When used in this way its performance is primarily determined by the magnitude of the external components used to apply the feedback. Figure 2.1 shows examples of the basic operational amplifier feedback configurations; all the circuits employ resistive feedback. It is suggested that the reader experimentally check the performance equations given, and calculate the range of signal values and component values for which the circuits operate satisfactorily.

Fig. 2.1 (a) Simple inverter, (b) inverting adder, (c) subtracting amplifier, and (d) non-inverting amplifier

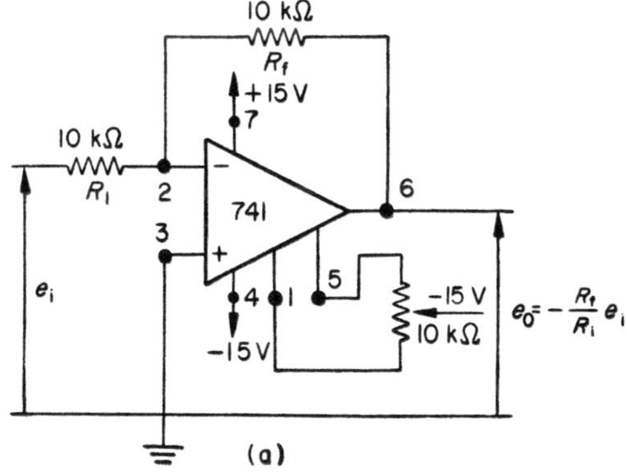

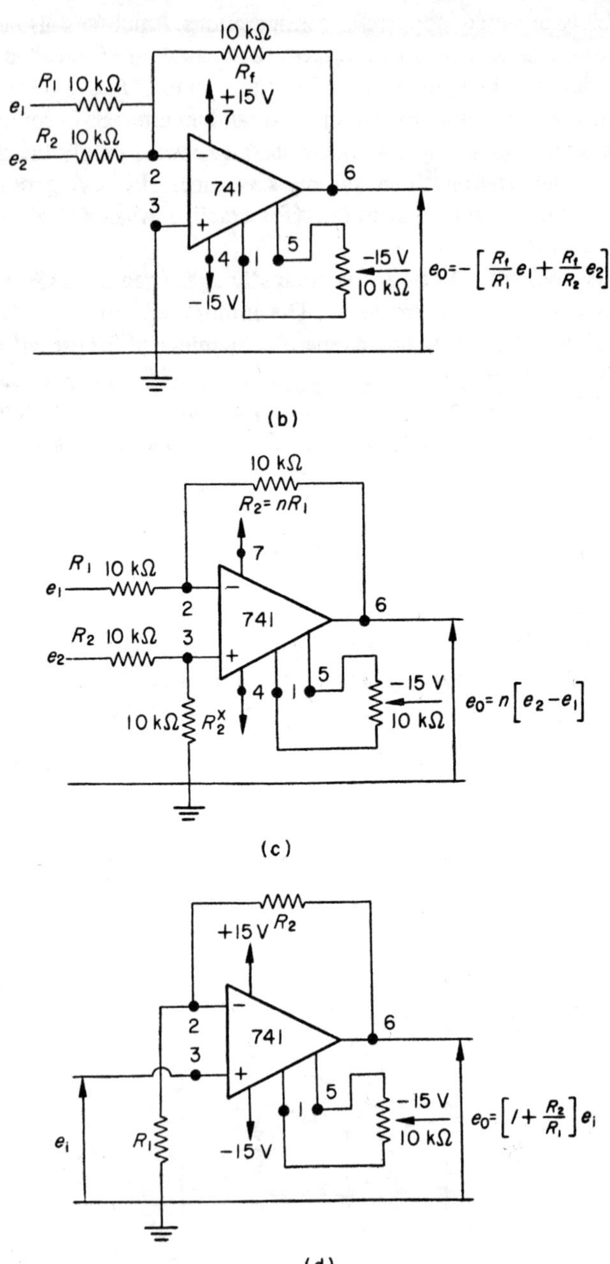
10 kΩ
R_f
R_1 10 kΩ
e_1
R_2 10 kΩ
e_2
+15 V
7
2
3
741
+
6
5
4
1
−15 V
−15 V
10 kΩ
$e_0 = -\left[\dfrac{R_f}{R_1}e_1 + \dfrac{R_f}{R_2}e_2\right]$
(b)
10 kΩ
$R_2 = nR_1$
R_1 10 kΩ
e_1
R_2 10 kΩ
e_2
7
2
3
741
+
6
5
4
1
10 kΩ R_2^x
−15 V
10 kΩ
$e_0 = n\left[e_2 - e_1\right]$
(c)
R_2
+15 V
2
3
741
+
6
5
1
−15 V
−15 V
10 kΩ
e_i
R_1
$e_0 = \left[1 + \dfrac{R_2}{R_1}\right]e_i$
(d)

Before taking readings from the circuits the input signal points should be earthed and the offset voltage potentiometer adjusted for zero d.c. voltage at the output of the amplifier. Input signals, both d.c. and a.c., may then be applied and input and output signals measured and recorded. Readings may be repeated using different component values in the circuits. An oscilloscope provides a convenient method of monitoring and measuring a.c. signals. A typical oscilloscope display obtained when the circuit of figure 2.1(b) is used to add together a triangular wave and a square wave is shown in figure 2.2. In the circuit of figure 2.1(c) a subtraction of equal inputs to give zero output requires a trimming of resistor values. The adjustment is performed by connecting together the two input points in the circuit and applying a sinusoidal signal. R_2^x is then adjusted for an a.c. null at the output.

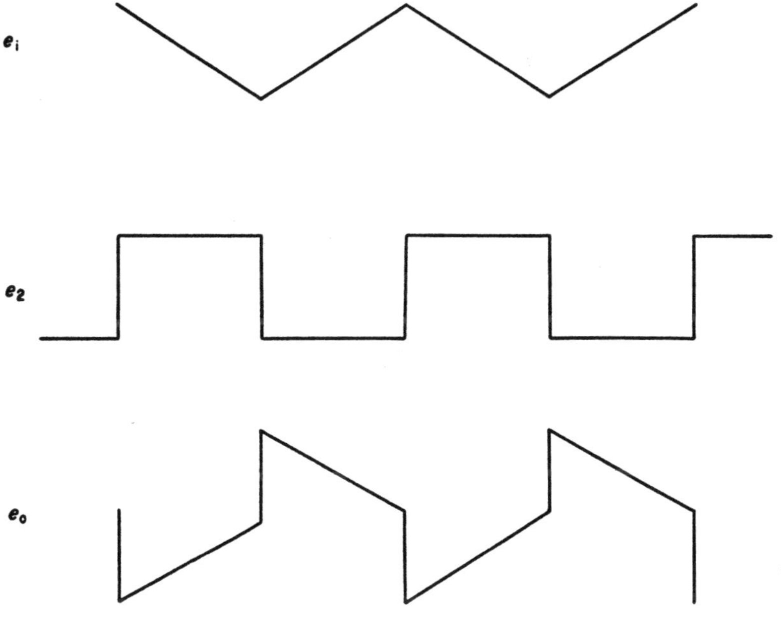

Fig. 2.2　　Summation of triangular and square wave

2.2.1 *Closed Loop Gain and Bandwidth.* (see appendix)

The bandwidth of a feedback amplifier is dependent upon the magnitude of the closed loop gain. The effect is conveniently investigated for the follower circuit of figure 2.1(d). A typical set of experimental results is tabulated in figure 2.3. In performing the test the input signal should be adjusted so that the output signal amplitude is never greater than a few hundred millivolts, if this is not done the effects of slew rate limiting at the amplifier output will prevent the measurement of true small signal bandwidths. Note also that a d.c. path to earth must be provided to allow bias current to flow to the non-phase-inverting input terminal of the amplifier. If the signal source does not provide such a path a resistor must be connected between the non-phase-inverting input terminal and earth. The results of the experiment are shown plotted on a log graph in figure 2.3. Note that the asymptotic closed loop bandwidths lie on the 20 dB/decade slope which represents the amplifier open loop frequency response.

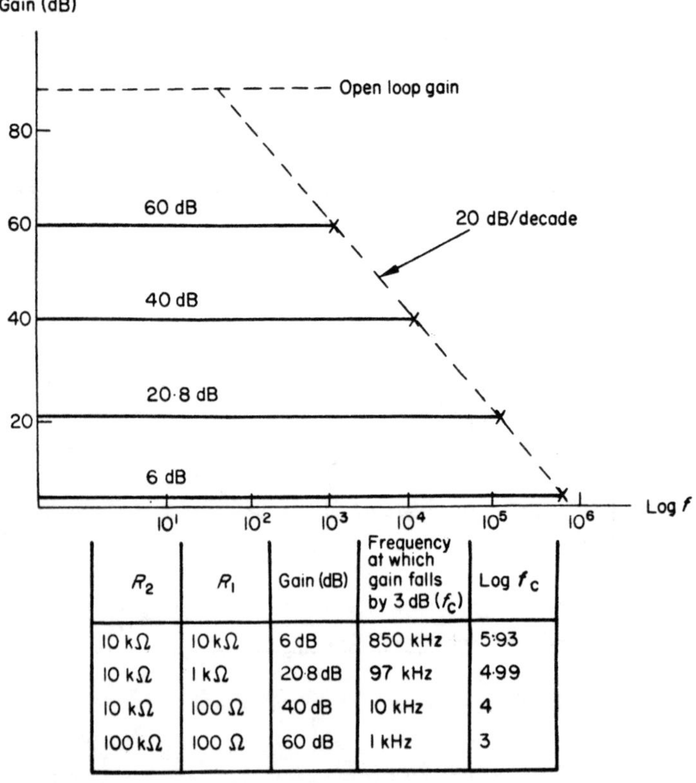

R_2	R_1	Gain (dB)	Frequency at which gain falls by 3 dB (f_c)	Log f_c
10 kΩ	10 kΩ	6 dB	850 kHz	5·93
10 kΩ	1 kΩ	20·8 dB	97 kHz	4·99
10 kΩ	100 Ω	40 dB	10 kHz	4
100 kΩ	100 Ω	60 dB	1 kHz	3

Fig. 2.3   Closed loop gain and bandwidth

2.2.2 *Current to Voltage Converter*

An operational amplifier may be used to measure current in two different ways.
The current may be converted into a voltage by passing it through a resistor and
the operational amplifier, connected as a follower, then used to amplify this
voltage. Alternatively the current may be injected directly into the summing
point of the amplifier connected in the inverting configuration. Under these
circumstances the current is forced to flow through the feedback resistor and the
output voltage of the amplifier takes on a value

$$e_0 = -I_{in} R_f$$

The current is converted into a voltage with a scaling factor R_f volts/amp.

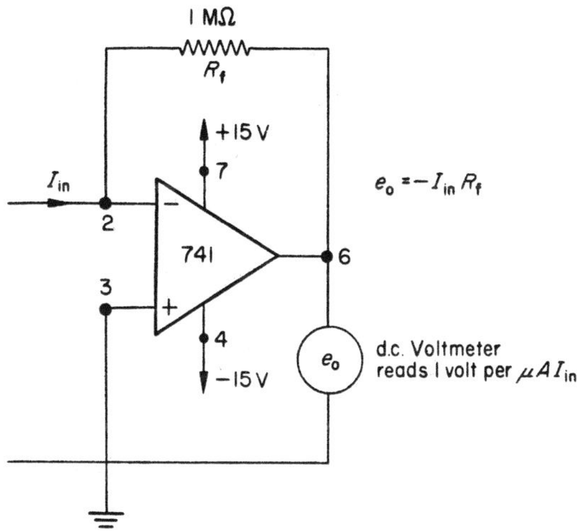

Fig. 2.4   Simple current to voltage converter

A simple circuit for a current to voltage converter is shown in figure 2.4. In
this circuit the effective input resistance at the phase inverting input terminal is
very small which means that a current measurement made with the circuit
introduces a negligible voltage drop in the measurement circuit. The simple
circuit gives a non-zero output voltage offset when the input current is zero, this
can be nulled for accurate measurements. If the circuit is to be supplied by
differing source impedances, or if R_f is to be changed (for a change of scale),
output offset should be nulled by separate balancing of both input voltage offset
and input bias current. A circuit which includes provision for these adjustments
is shown in figure 2.5. Input offset voltage is balanced first, pins 2 and 3 are

connected together with a 100 Ω resistor, and the 10 kΩ offset balance potentiometer is adjusted to make the amplifier output voltage zero. The connection between pins 2 and 3 is removed and with zero input current the bias current potentiometer is adjusted so that the amplifier output is again zero.

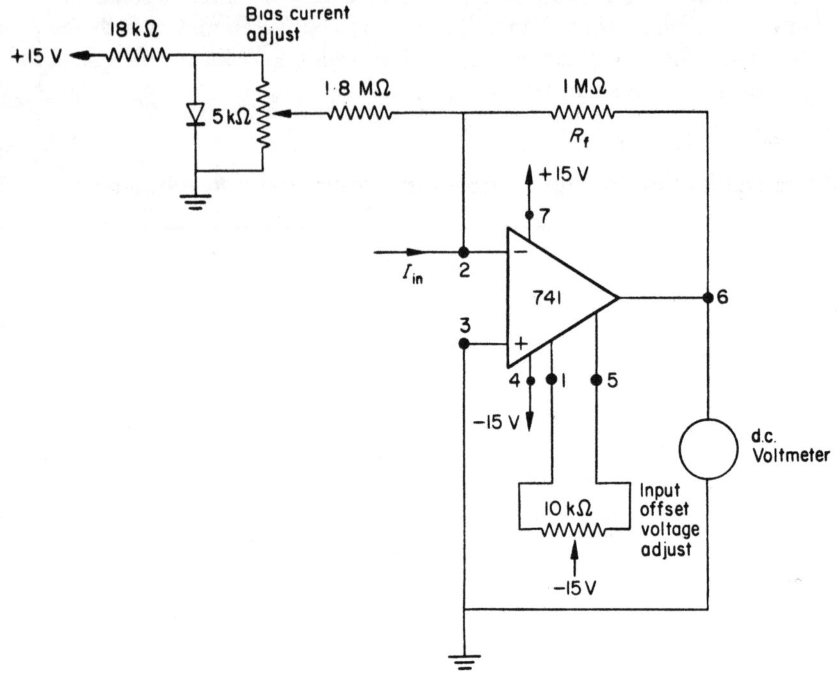

Fig. 2.5   Current to voltage converter with offset adjustment

2.3 Experiments with an Operational Integrator

An operational amplifier with negative feedback applied to it by means of a capacitor connected between the amplifier output terminal and the phase inverting input terminal may be used to perform the operation of integration. A circuit for a simple operational integrator is illustrated in figure 1.6. If the performance of the amplifier is assumed to be ideal the response of the circuit is described by the equation

$$e_0 = \frac{-1}{CR} \int e_i \, dt \qquad\qquad (2.1)$$

The time constant, $T = CR$, is called the characteristic time of the integrator. It is sometimes convenient to think of $1/T$ as the integrator 'gain' in terms of volts/second output per volt input.

Real amplifiers show departures from ideal behaviour, in particular, amplifier input offset voltage and bias current cause a continuous charging of the capacitor C in figure 2.6 even when the input voltage e_i is made zero. The output of the simple integrator thus drifts with time and the amplifier eventually saturates.

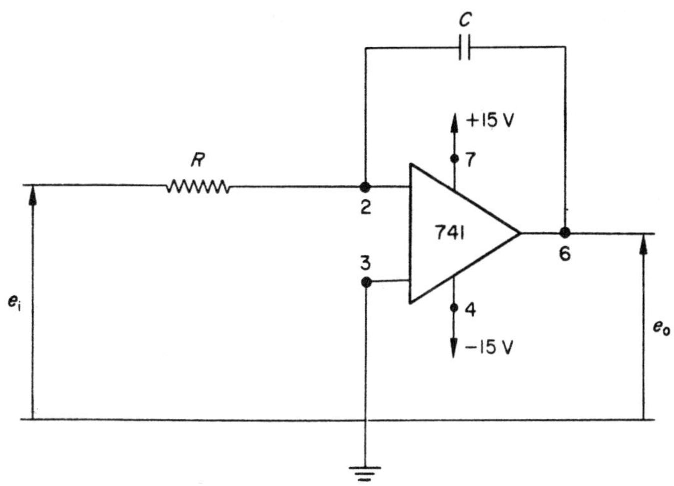

Fig. 2.6   Simple integrator

2.3.1 *Measurement of Integrator Drift*

The drift rate of the output voltage of the amplifier in figure 2.6 with e_i zero (resistor R connected to earth), is given by the relationship

$$\frac{de_o}{dt} = \pm \frac{V_{io}}{CR} + \frac{I_b^-}{C} \tag{2.2}$$

Where V_{io} is the amplifier input offset voltage and I_b^- is the bias current drawn by the phase inverting input terminal of the amplifier.

A practical test of the validity of equation 2.2 can be made by measuring drift rates for various values of capacitor C and resistor R. The following test procedure is suggested. The output of the integrator is applied to the d.c. coupled vertical amplifier of an oscilloscope, the input end of resistor R is earthed. The timebase of the oscilloscope is set to free run at a slow rate (say 1 div/sec). The integrator output is initially set to zero by shorting capacitor C, the short is removed and the drift rate is determined directly by observation and measurement of the slope of the oscilloscope trace. This slope may be positive or negative, it gives the sign of the drift rate which is significant and should therefore be noted. If an oscilloscope with slow sweep speeds is not available the output of the amplifier may be measured with a centre zero voltmeter. Drift rates may then be found by measuring the change of output voltage that takes place in a measured period of time. The measurement period should not be so long that it allows the amplifier to drift into saturation.

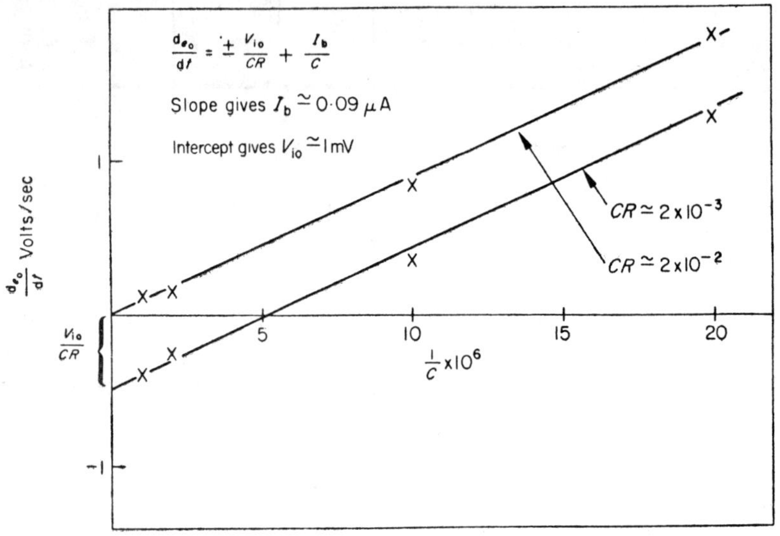

	R	C	$\frac{1}{C}$	$\frac{de_o}{dt}$
$CR \simeq 2 \times 10^{-3}$	39 kΩ	0.05 μF	2×10^7	+1.3
	22 kΩ	0.1 μF	10^7	+0.35
	3.9 kΩ	0.5 μF	2×10^6	−0.28
	2.2 kΩ	1 μF	10^6	−0.4
$CR \simeq 2 \times 10^{-2}$	390 kΩ	0.05 μF	2×10^7	+1.8
	220 kΩ	0.1 μF	10^7	+0.84
	27 kΩ	0.5 μF	2×10^6	+0.16
	22 kΩ	1 μF	10^6	+0.1

Fig. 2.7   Measurement of integrator drift

A set of experimentally obtained readings are tabulated in figure 2.7. Note that the readings obtained in this experiment are dependent upon the values of the input offset voltage and bias current of the particular amplifier used. These parameters are subject to a considerable spread and results obtained are therefore likely to show marked differences for different amplifiers. In figure 2.7 graphs of de_0/dt against $1/C$ are plotted for fixed values of CR. Inspection of equation 2.2 shows that the slopes of the graphs give the bias current, I_b^-, and that the intercept on the de_0/dt axis may be used to estimate a value for the input offset voltage of the amplifier, V_{io}.

The results illustrate the general point that for a fixed CR value, minimum drift is likely to be obtained by using as large a value as possible for the capacitor C. In fact the magnitude of C that it is practicable to use is dependent upon the leakage of the capacitor since capacitor leakage current introduces another source of drift. Also the larger the value of C the smaller is the value required for R. A small value of R may unduly load the signal source which is supplying the input signal to the integrator.

2.3.2 *Examination of Integrator Action*

The action of a simple integrator can be investigated using the circuit illustrated in figure 2.8. The integrator input is connected to earth and the output is set to zero by momentarily closing the reset switch. The offset balance potentiometer is now adjusted for minimum output drift; it should be possible to obtain almost zero drift, over the short term, by means of this adjustment.

The circuit performs the operation of integration on an input signal in the following manner. Suppose a -1 volt d.c. signal to be applied to the input resistor R (1 MΩ), this will draw a current of 1 μA, from capacitor C (1 μF). In order to produce this current a continuously rising voltage must be generated at the output with a rate of rise equal to 1 volt/second. This rate of rise continues until the input voltage is changed or the amplifier output reaches saturation. If the input is switched to zero the output voltage of the amplifier ideally remains at the level reached when the input was removed. If now a $+1$ volt d.c. signal is applied to R it causes a charging current of opposite polarity in capacitor C and consequently a continuously falling voltage must be generated at the amplifier output with a rate of fall of 1 volt/second.

The integrator output level at the end of any period of time is equal to the sum of the products of each voltage times the period of each applied voltage divided by $-RC$. The change in output level during any time period is thus proportional to the area under the input voltage/time graph for that period of time.

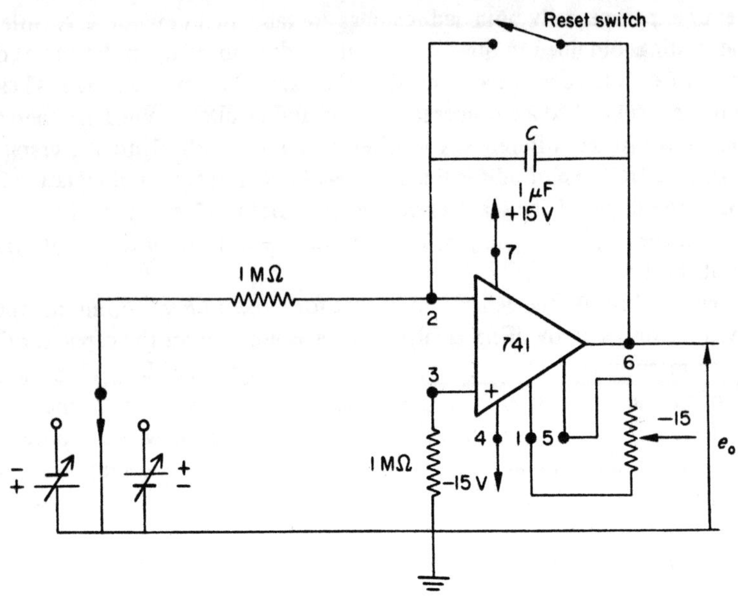

Fig. 2.8   Examination of integrator action

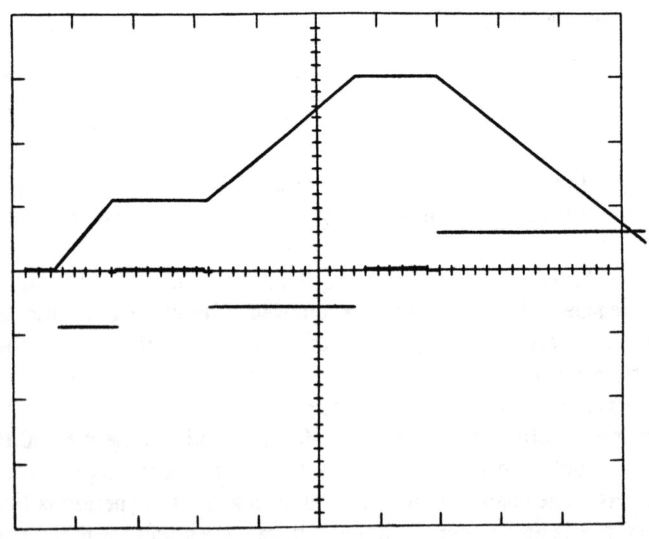

**Fig. 2.9   Integrator response; horizontal 5s/division; upper trace, integrator
output 2V/division; lower trace, integrator input 1V/ division**

The action of the integrator can be verified experimentally by switching the input to known d.c. levels for measured periods of time whilst at the same time monitoring the amplifier output voltage. Integrator input and output voltages as the input is switched between different d.c. levels are illustrated by the oscilloscope traces shown in figure 2.9.

2.3.3 *An Integrator used to produce a Linear Staircase Waveform*

An operational integrator may be used to linearise the output of a simple diode pump circuit and thus allow the generation of a linear staircase waveform. A circuit which demonstrates this action is illustrated in figure 2.10. In the circuit a constant amplitude square wave of amplitude V_{in} (approximately 10 V say), and frequency f (say 400 Hz.), is applied to the capacitor C_1. Capacitor C_1 charges through diode D_1 on the negative-going part of each input wave. On the positive-going part of each square wave, C_1 discharges through diode D_2 thus transferring a quantity of charge C_1 V_{in} to the integrating capacitor C_2 and causing a step decrease in the output voltage of the amplifier. The amplitude of the output step is equal to C_1 V_{in}/C_2 and the output falls in successive steps until the amplifier saturates or its output is reset to zero in some way. In the circuit of figure 2.10 a UJT is used to reset the integrator. The output voltage of

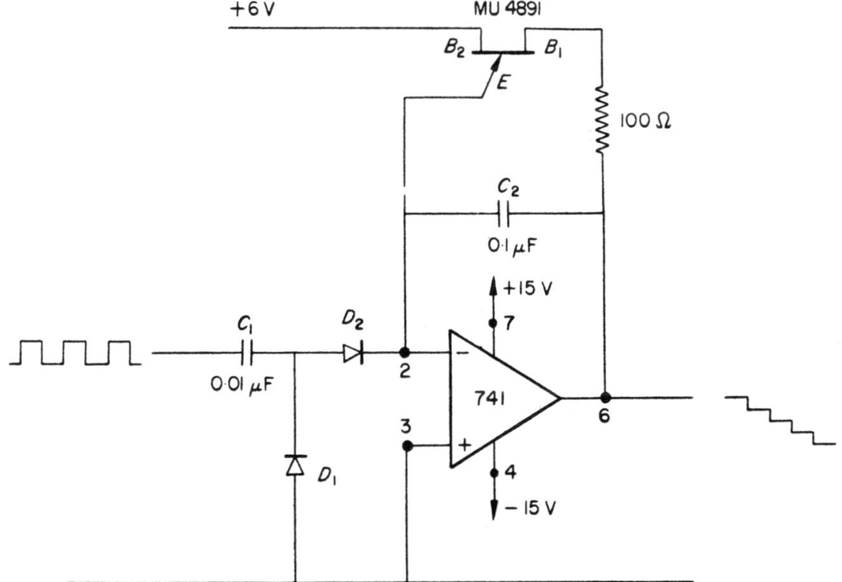

Fig. 2.10   Linear staircase generator

the amplifier is applied between B_1 and emitter of the UJT and the capacitor is discharged when V_{EB} becomes equal to $\eta\, V_{B2\,B1}$; η is the intrinsic stand off ratio of the UJT. The integrator output level at which reset takes place is thus controlled by the voltage applied to the UJT B_2 terminal. Values of V_{B2} in the range 2–6V should be found suitable.

Typical input and output waveforms for the staircase generator are shown in figure 2.11. It is suggested that input frequency, input amplitude, unijunction B_2 voltage and finally capacitor values be changed in turn, the effects of each change being noted and explained in terms of the action of the circuit. If a very low frequency input square wave is used, integrator drift will cause the output voltage to change appreciably between steps. The effect can be remedied by connecting the usual 10 kΩ offset potentiometer to the amplifier between pins 1 and 5, 'Step flatness' can then be obtained by adjustment of this potentiometer.

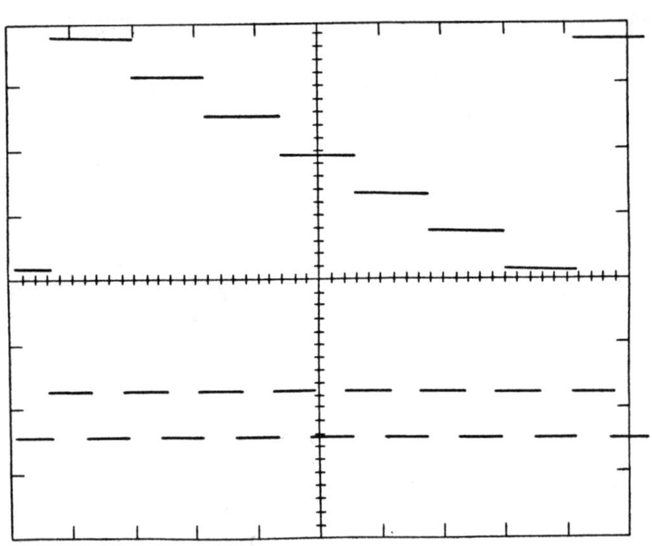

Fig. 2.11   Linear staircase waveform; horizontal 2ms/division; output staircase 1V/division; input square wave 10V/division

2.3.4 *Frequency to Voltage Conversion*

The staircase generator circuit of figure 2.10 can be modified to make the output of the amplifier a direct voltage of magnitude proportional to the frequency of a constant amplitude input square wave. The modification consists simply of replacing the UJT in the circuit with a resistor R connected in parallel with the capacitor C_2. With this modification the charge transferred per second through diode D_2, $f\,C_1\,V_{in}$, produces an average current, $f\,C_1\,V_{in}$, through the resistor R and the amplifier gives an output voltage

$$e_0 = -f\,C_1\,V_{in}\,R \tag{2.3}$$

It is suggested that measurements be taken in order to check the validity of equation 2.3. With resistor R given a value of say $100\ \text{k}\Omega$, the frequency of the input square wave should be increased in steps whilst maintaining its amplitude constant. The output voltage for each frequency should be measured and recorded. A graph of output voltage against frequency should then give a straight line of slope equal to $C_1\,V_{in}\,R$. The experiment may be repeated for different component values.

2.4 Experiments with an Operational Differentiator

An operational amplifier with negative feedback applied via a resistor connected between output terminal and phase inverting input terminal performs the operation of differentiation on a signal applied to the phase inverting input terminal through a capacitor. A simple differentiator circuit is illustrated in figure 2.12. The input current to the amplifier summing point is proportional to the rate of change of the input voltage applied to the capacitor. Feedback causes the input current to flow through the feedback resistor and the output voltage of the amplifier thus takes on a value which is proportional to the rate of change of the input voltage. If the performance of the amplifier is assumed to be ideal the response of the circuit is described by the equation

$$e_0 = -C R\,\frac{de_i}{dt} \tag{2.4}$$

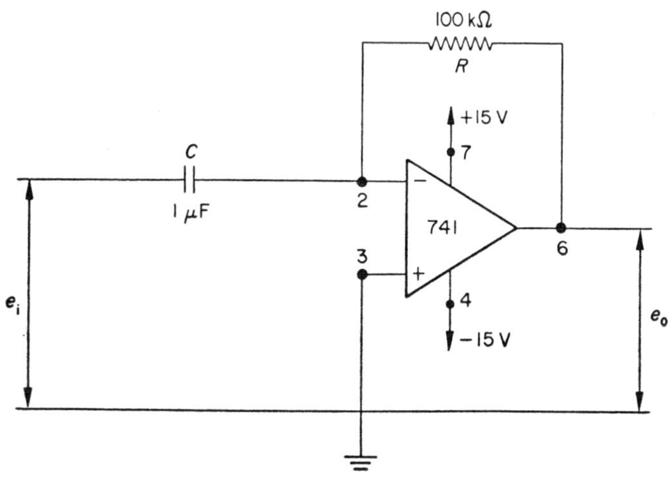

Fig. 2.12   Simple differentiator

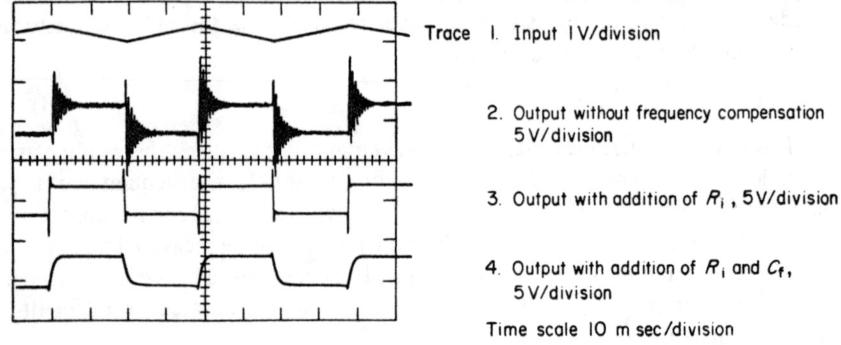

Fig. 2.13 Trace 1. Input 1V/division; 2. Output without frequency compensation 5V/division; 3. Output with addition of R_i 5V/division; 4. Output with addition of R_i and C_f 5V/division; time scale 10ms/division

2.4.1 *Frequency Compensation of a Differentiator*

A low frequency linear sawtooth wave is a convenient input test signal for examining the action of the simple differentiator. Typical input and output waveforms are shown in figure 2.13. In trace 2 the output voltage is seen to overshoot and ring in response to the sudden change in the slope of the input wave. If the closed loop sinusoidal frequency response of the circuit is measured it will be found to show a marked peaking at this ringing frequency. The effect arises because of phase shift with frequency around the feedback loop; certain frequencies undergo a phase shift approaching 180° when fed back to the amplifier input. The feedback is no longer negative for these frequencies and the closed loop gain thus shows its marked peaking. Two factors contribute to the phase shift, phase lag caused by the fall off in the amplifier open loop response at frequencies outside its 3 dB bandwidth limit and an additional phase lag caused by the input capacitor C. Both factors can individually introduce a phase lag approaching 90°, giving a total phase lag which can approach 180°.

Overshoot and ringing can be prevented by adding either a resistor R_i in series with the capacitor C or a capacitor C_f in parallel with resistor R. Alternatively both R_i and C_f may be added to the circuit. The circuit in figure 2.14 shows the simple differentiator with the addition of frequency compensating components and traces 3 and 4 of figure 2.13 show the effects of the frequency compensating components on the output waveform. The frequency compensating components used in the test circuit are *not* optimum values. Readers should refer to the appendix for considerations governing the choice of values.

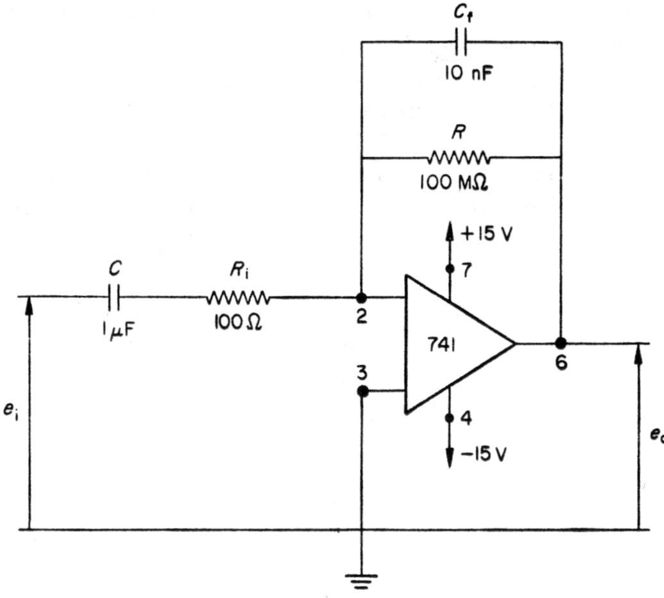

Fig. 2.14   Differentiator with frequency compensation

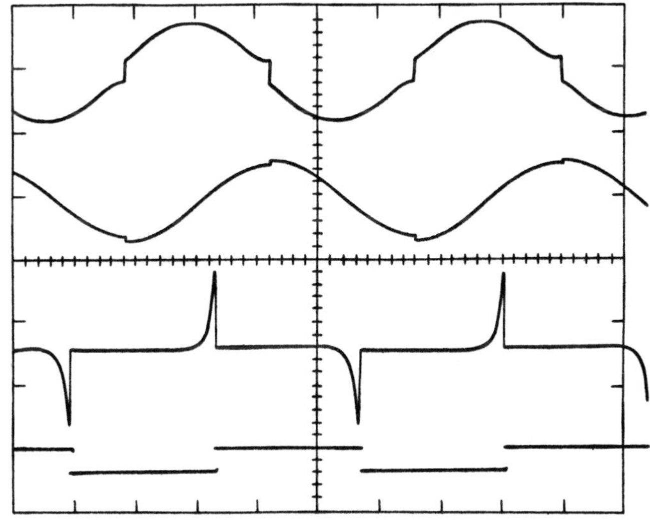

Fig. 2.15   Response of differentiator to different input waveforms

The validity of equation 2.4 for the frequency compensated differentiator may be checked by varying the amplitude and frequency of the input sawtooth voltage and measuring the output wave for each input signal. It is also instructive to observe the response of the differentiator to a variety of different input waveforms. The traces in figure 2.15 show input and output waveforms for a square wave input and for a sinusoidal input. Note that the input sinusoid in figure 2.15 shows a slight discontinuity at its peak value, the presence of this discontinuity is far more apparent in the differentiated waveform.

2.4.2 *An Application of a Differentiator*

The transient response of a control or feedback system is often observed in order to investigate the stability of the system. In such observations it is often difficult to distinguish effects due to non-linear elements. If the transient signal is plotted against the differentiated transient.non-linear effects are more readily observed.

A circuit which can be used to illustrate the principle of this type of application is shown in figure 2.16. The effect of resistive and diode damping on an LCR circuit is first compared in a transient response display, figure 2.17. Values are chosen so that the diode does not conduct heavily and it is difficult to distinguish the effect of diode non-linearity.

The transient is now used to produce the horizontal deflection and the differentiated transient is used to produce the vertical deflection of the oscilloscope display. The traces obtained are shown in figure 2.18. The effect of the diode non-linearity is clearly apparent in the lower trace. Note that the input square wave used to excite the LCR circuit is applied to the oscilloscope to blank off the display of the negative-going transient.

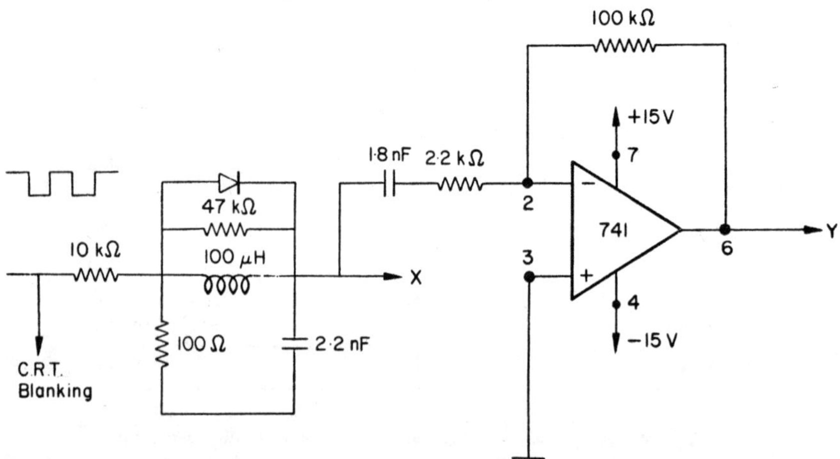

Fig. 2.16   Obtaining a plot of $\frac{de}{dt}/e$ for a transient

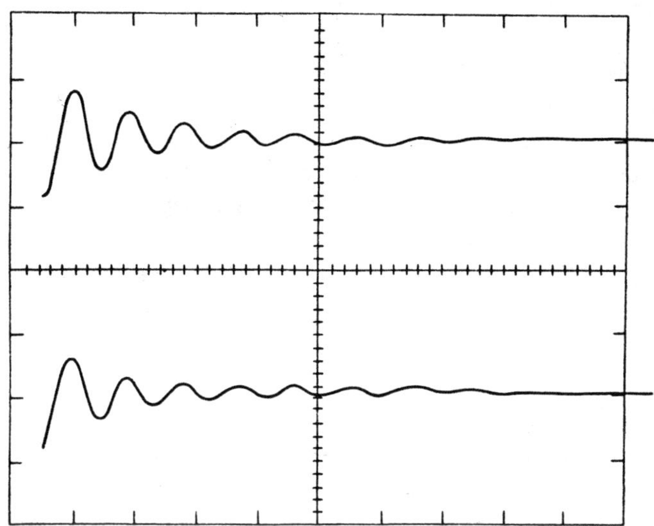

Fig. 2.17   Transient response of LCR circuit; upper trace, resistive damping; lower trace, diode damping

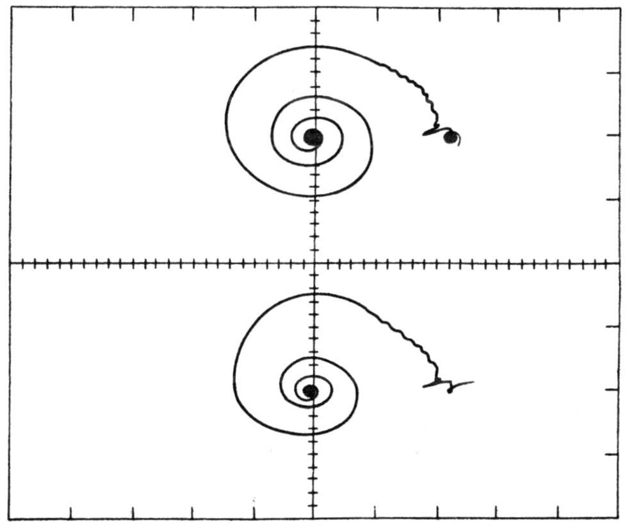

Fig. 2.18   Plot of $\frac{de}{dt}$ /e; upper trace, resistive damping; lower trace diode damping

Exercises

General Notes for Guidance (see also appendix)

In negative feedback circuit configurations the open loop gain of the amplifier used in the circuit controls the closed loop gain errors and the small signal open loop frequency response characteristics of the amplifier directly influence the small signal closed loop bandwidth and the closed **loop** stability.

Gain Terminology

It is important to distinguish between the **several** 'gain' terms which are often used when discussing operational amplifier feedback circuits.

Open Loop Gain

A_{OL}, may be defined as the ratio of a change of output voltage to the change in the input voltage which is applied directly to the amplifier input terminals.

The other gains are dependent upon both the amplifier and the circuit in which it is used and are controlled by the feedback fraction β.

The feedback fraction β is the fraction of the amplifier output voltage which is returned to the input. It is a function of the entire circuit from output back to input, including both designed and stray circuit elements and the input impedance of the amplifier.

Loop Gain, βA_{OL} is the total gain in the closed loop signal path through the amplifier and back to the amplifier input via the feedback network. The magnitude of the loop gain is of prime importance in determining how closely circuit performance approaches the ideal. The magnitude/phase relationships of βA_{OL} control closed loop stability.

Closed Loop Gain is the gain for signal voltages connected directly to the input terminals of the amplifier. The closed loop gain for an ideal amplifier circuit is $1/\beta$ and for a practical circuit is

$$\frac{1}{\beta} \qquad \left[\frac{1}{1 + \dfrac{1}{\beta \, A_{OL}}} \right]$$

The quantity

$$\left[\frac{1}{1 + \dfrac{1}{\beta \, A_{OL}}} \right]$$

is called the gain error factor

The amount by which this factor differs from unity represents the gain error (usually expressed as a percentage).

Signal Gain is the closed loop transfer relationship between the output and any signal input to an operational amplifier circuit.

Care should be taken to avoid confusion between closed loop gain and signal gain. In some circuits, the follower for example, the two gains are identical. Reference to the circuit shown in figure E.2.1 should clarify the distinction between the two types of gain.

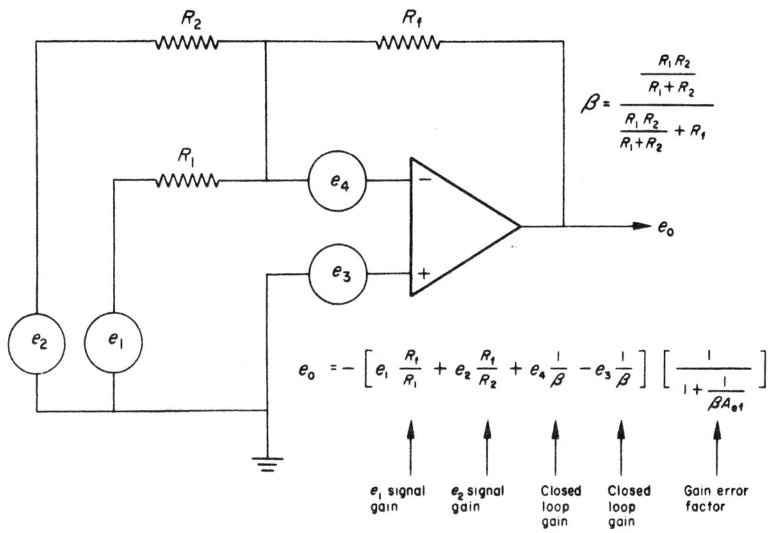

Fig. E.2.1 Difference between signal gain and closed loop gain

It is worth noting that amplifier offset voltage, drift and noise, as specified, are referred to the input and are represented in terms of equivalent generators at the input connected at the position occupied by the signal source e_3 or e_4 in figure E2.1. They appear at the output multiplied by the closed loop gain which for βA_{OL} much greater than 1 is equal to $1/\beta$. It is for this reason that $1/\beta$ is referred to as the 'noise' gain. An amplifier input offset voltage e_{os} becomes $1/\beta$ e_{os} at the output but when referred to the signal inputs in figure E.2.1 it becomes

$$e_{os} \; \frac{1}{\beta} \; \frac{R_1}{R_f}$$

at the e_1 input and

$$e_{os} \; \frac{1}{\beta} \; \frac{R_2}{R_f}$$

at the e_2 input.

The General Offset and Drift Case

An appreciation of the difference between closed loop gain and signal gain, may be used to set out a general method for characterising and evaluating offset and drift errors.

Express all offsets in terms of equivalent voltages connected directly to the input terminals. Thus $I_{b\mp}$ applies a voltage $-I_{b-} \, R_{source}$ to the inverting input terminal and I_{b+} applies a voltage $-I_{b+} R_{source_+}$ to the non-inverting input terminal.

R_{source_-} and R_{source_+} represent the effective source resistances connected at the inverting and non-inverting input terminals respectively. They represent the parallel combination of all resistive paths to ground, including, in the case of R_{source_+}, the path through any feedback resistor and the amplifier output resistance to ground.

V_{io} is directly applied to the input terminal so we may represent the total equivalent input offset voltage as

$$E_{os} = \pm \, V_{io} + I_{b-} \, R_{source_-} - I_{b+} \, R_{source_+}$$

Drift in the total equivalent input offset voltage is obtained by substituting values of the drift coefficients of I_b and V_{io}.

Graphs showing the dependence of E_{os} drift on source resistance are given in some amplifier data sheets. E_{os} appears at the output multiplied by the 'noise gain' $1/\beta$. The resultant error may be referred to any signal input simply by dividing by the signal gain associated with that input.

Amplifier Parameters

The characteristics of the operational amplifier to be used in the numerical examples are

    Open loop gain 80 dB

    First Order frequency response determined by

$$A_{OL\,(j\omega)} = \frac{10^4}{1 + j\,10^{-3}\,\omega}$$

$$\text{CMRR} = 80\text{ dB} \quad I_B = 500\text{ nA}$$

$$\frac{\Delta I_B}{\Delta T} = 1\text{ nA/}^\circ\text{C}, \quad I_{io} = 50\text{ nA}, \quad \frac{\Delta I_{io}}{\Delta T} = 0.1\text{ nA/}^\circ\text{C}$$

$$V_{io} = 2\text{ mV}, \quad \frac{\Delta V_{io}}{\Delta T} = 10\ \mu\text{V/}^\circ\text{C}$$

Exercises 2

2.1   A simple inverting amplifier (figure 2.1a) has $R_1 = 10\text{ k}\Omega$ $R_f = 1\text{ M}\Omega$.

(a)   What is the closed loop bandwidth of the circuit?

(b)   What is the ideal closed loop signal gain and the gain error at zero frequency?

(c)   If no offset balance is employed, what is the output offset voltage?

(d)   What is the change in output offset voltage to be expected from a temperature change of 10°C?

(e)   Assuming that the initial value of the output offset voltage is balanced out, what is the minimum input signal that can be amplified with less than 1 per cent error due to a temperature change of 10°C?

(f)   In order to reduce the offset error due to amplifier bias current a resistor R_c is connected between the non-inverting input terminal of the amplifier and earth. What value of resistor should be used?

Repeat parts c, d and e assuming that the resistor R_c is connected ir the circuit.

2.2   Sketch a circuit configuration which will produce an output signal related to the value of three input signals e_1, e_2, e_3 by the equation

$$e_0 = -(e_1 + 4\,e_2 + e_3)$$

2.3   A summing amplifier configuration is connected up (figure 2.1b) using th- following resistance values

$$R_f = 500\text{ k}\Omega, \quad R_1 = 20\text{ k}\Omega, \quad R_2 = 5\text{ k}\Omega$$

(a)   What is the value of the feedback fraction β?

(b)   What is the closed loop bandwidth?

(c)   What are the ideal values of the two closed loop signal gains?

(d)   What is the gain error at zero frequency?

(e) What is the output offset voltage with no offset balance?

(f) What change in output offset is to be expected for a $10°C$ change in temperature?

(g) Assuming that the initial offset voltage is balanced out what is the minimum signal which can be amplified with less than 1 per cent error due to a temperature change of $10°C$:

 (i) at the e_1 input

 (ii) at the e_2 input

(h) In order to reduce the offset error due to amplifier bias current, a resistor R_c is connected between the non-inverting input terminal of the amplifier and earth. What value of resistor should be used?

 Repeat parts e, f and g assuming that the resistor R_c is connected in the circuit.

2.4 What is the percentage error introduced in the non-inverting amplifier configuration (figure 2.1d) due to the non-infinite CMRR of the amplifier?

2.5 An operational amplifier is to be used as a simple current to voltage converter to be supplied by a current source of internal resistance 1 MΩ, a scaling factor of 0.2 volts per microamp is required. Sketch the circuit.

(a) If no offset balance is used what is the minimum current that can be measured with less than 1 per cent error?

(b) If initial offsets are balanced what is the minimum current that can be measured with less than 1 per cent error for a temperature change of $10°C$?

2.6 A simple operational integrator (figure 2.6) uses $C = 0.5\ \mu F$, $R = 1\ M\Omega$. The output of the integrator is initially zero, the input signal to the integrator is switched to the following values after successive time intervals.

Time t (seconds)	0	3	8	10	12	18
Input Signal e_i	$+1V$	$+0.1V$	$-4V$	$0V$	$+0.5V$	$0V$

Sketch the time variation of the integrator output voltage. What is the value of the integrator output voltage after 19 seconds? Neglect integrator drift.

2.7 What is the drift rate of the simple integrator in question 2.6?

(a) with no drift balance.

(b) if the drift is initially balanced and the temperature changes by $10°C$.

2.8 An operational integrator with feedback capacitor $C_2 = 1\ \mu F$ is to be used to produce a positive-going staircase waveform from a square wave signal with peak to peak value 10 volt and frequency 100 Hz. The staircase is to be set to zero when the output reaches 10 volts after 100 steps. Sketch the circuit. What 'sag' is to be expected in the steps if drift is not balanced?

2.9 The simple differentiator circuit of figure 2.12 is used with component values, $C = 0.1\ \mu F$, $R = 100\ k\Omega$. In what way do you expect the circuit performance to differ from that of an ideal differentiator? What will be the time response of the output signal in response to a sudden change in input signal?

A resistance of 1 $k\Omega$ is connected in series with the input capacitor C. In what way does this affect the circuit performance?

At what signal frequency will the output of the modified circuit be in error by 3 dB? Illustrate your answer by an appropriate Bode plot.

3. Operational Amplifier Circuits with a Non–linear Response

An amplifier is normally required to give an output signal which faithfully reproduces the waveform of the input signal applied to it and if it fails to do this it is said to introduce distortion. Distortionless amplification demands the use of an amplifier for which the relationship between input and output signals is linear. An operational amplifier with negative feedback applied to it by means of linear components gives linear amplification. There are certain specialised amplifier applications however, in which an amplifier is required to process its input signal in a defined non-linear manner. Examples of these non-linear applications may be found in analogue computing circuits, in circuits designed to generate non-linear functions and in circuits used to linearise transducer characteristics in medical, industrial and process control equipment. In a later section of this book, which deals with function generator systems, a non-linear amplifier is used to shape a triangular wave into a sinusoidal wave.

This chapter is mainly concerned with the presentation of practical circuits for experimentally investigating the ways in which operational amplifiers can be used to give non-linear amplification. The first circuit presented demonstrates a way in which a defined non-linear response can be synthesised by a series of linear approximations, the remainder of the chapter deals with operational amplifier circuits for log and antilog conversions and with some applications of these circuits. The material is intended as an experimental supplement and not as a duplication of material in a book previously published by the author. The reader is referred to this book for a fuller discussion of non-linear amplification together with further reading suggestions[1].

3.1 Straight Line Approximated Non-Linear Response

Straight line approximated non-linear functions can be generated by using an operational amplifier to sum a series of currents, the current sums being used to define the slopes of the linear approximations. A circuit which can be used to experimentally investigate this technique is shown in figure 3.1. In this circuit the individual current components in a current sum are determined by the choice

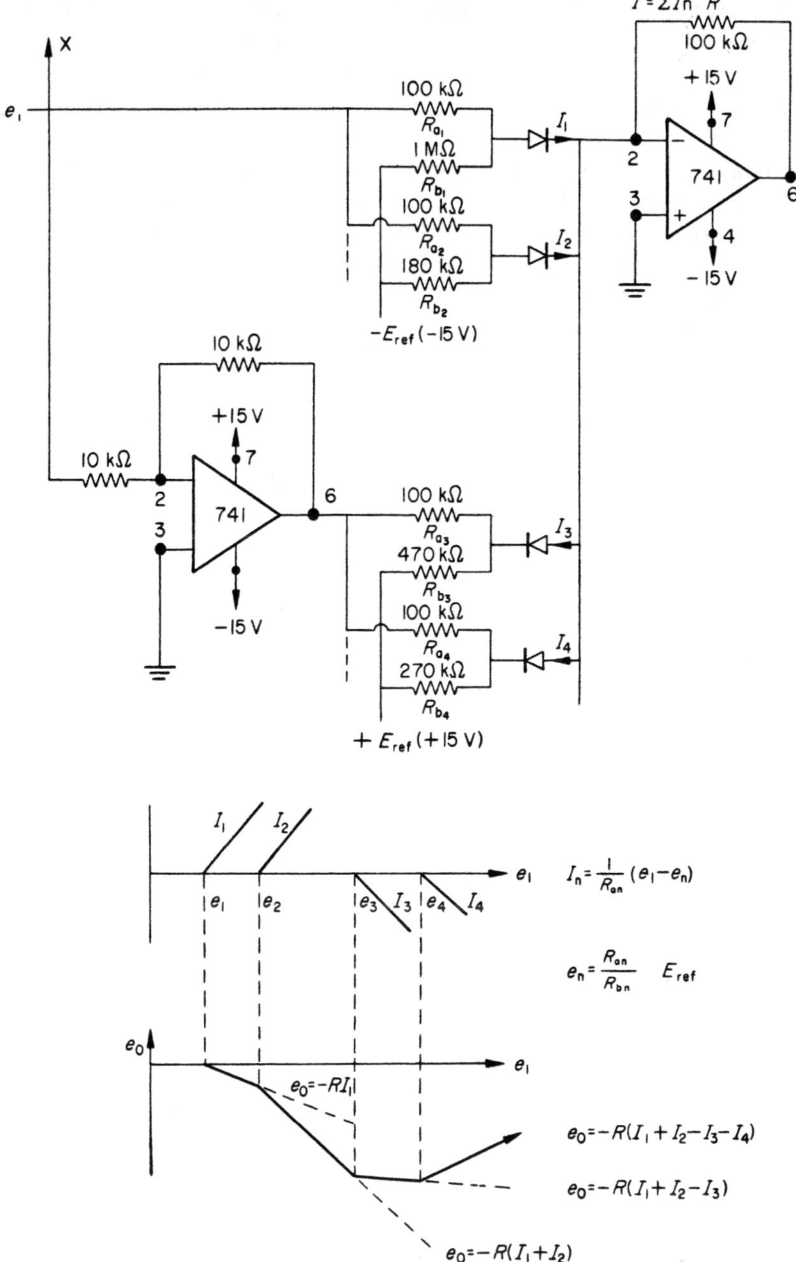

Fig. 3.1   Synthesis of a non-linear function

of input resistor values and the break points at which the changes of slope occur are set by a reference voltage, a diode and a resistive dividing network.

An oscilloscope display of the circuit response can be obtained by applying a low frequency sinusoidal signal to the input circuit point. This signal is used to provide horizontal deflection and the amplifier output provides the vertical deflection for the display. Typical waveforms obtained in this way are shown in figure 3.2. In these traces the first graticule line at the left represents the horizontal zero reference; the amplifier is seen to give zero output for negative values of the input signal. The following component values were used for the upper and lower traces respectively: R 100 K.; R_{a1} 150 kΩ, 100 kΩ; R_{b1} 1.8 MΩ, 1 MΩ; R_{a2} 100 kΩ, 100 kΩ; R_{b2} 180 kΩ, 180 kΩ; R_{a3} 100 kΩ, 100 kΩ; R_{b3} 470 kΩ, 470 kΩ; R_{a4} 120 kΩ, 100 kΩ; R_{b4} 270 kΩ, 270 kΩ.

It is instructive to change component values and to note the effect of such changes on slopes and break points. Observed results should be compared with theoretically expected values. Discrepancies are to be expected between measured values and values predicted by the simple theoretical treatment presented in figure 3.1. This treatment assumes ideal diode characteristics and neglects diode forward voltage drops. Additional input networks can be added to the circuit in order to introduce additional straight line segments and negative input signals can be processed by adding extra input networks with diode and reference voltage polarities reversed.

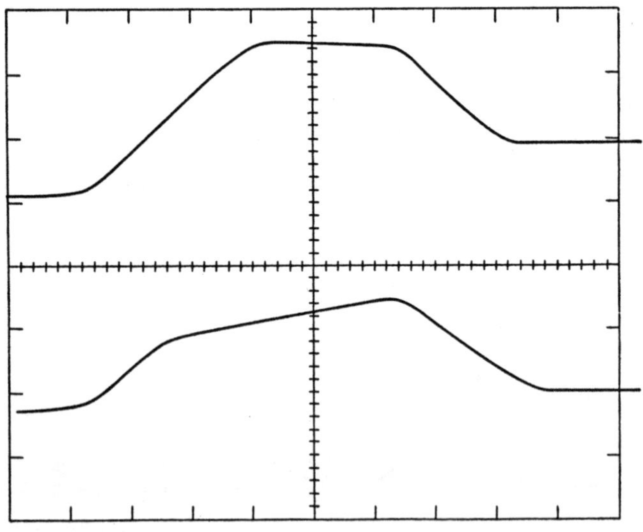

Fig. 3.2   Synthesised non-linear response

3.2 Operational Amplifier Transistor Feedback Circuits for Logarithmic Conversion

Bipolar transistors, when operated under appropriate conditions, are found to behave logarithmically and an operational amplifier with a transistor connected in its feedback path may be used to perform logarthmic conversions[2]. Logarithmic converters using this principle assume that the transistor follows a characteristic described by the equation

$$V_{EB} = -E_0 \log_{10} \frac{I_c}{I_0} \qquad\qquad (3.1)$$

I_c is the collecter current in amps

I_0 is a constant at constant temperature with a value typically of order 10^{-12} A

E_0 is a constant at constant temperature, its value is approximately 60 mV at 27°C.

V_{EB} is the emitter base voltage of the transistor.

Provided that their collecter base voltage is held at zero, certain silicon planar transistor types are found to obey this equation accurately over a wide range of collecter current values. Because of the temperature dependence of the terms I_0 and E_0, simple log converters using single transistors give accurate logarithmic conversion only if the temperature is held constant. Temperature dependence may be considerably reduced by employing a circuit in which two transistors are connected in such a way as to balance out the effects of their temperature dependent parameters, the arrangement requires the use of two operational amplifiers. Experimental circuits for investigating the action of both simple log converters and temperature compensated converters are now described.

3.2.1 *A Simple Logarithmic Converter*

A circuit suitable for investigating the behaviour of a simple logarithmic converter is shown in figure 3.3. When a positive input voltage is applied to the circuit negative feedback is applied to the amplifier because of current flow through the diode connected transistor T_1. Diode connection of the transistor ensures that its collecter base voltage remains zero but restricts the logging range to values of input current larger than the collecter current value at which the current gain of the transistor begins to fall appreciably. Diode connected type 2N3707 transistors have been found to give accurate log conversion for currents down to 10^{-8} A. and below. Provided that the common emitter current gain of the

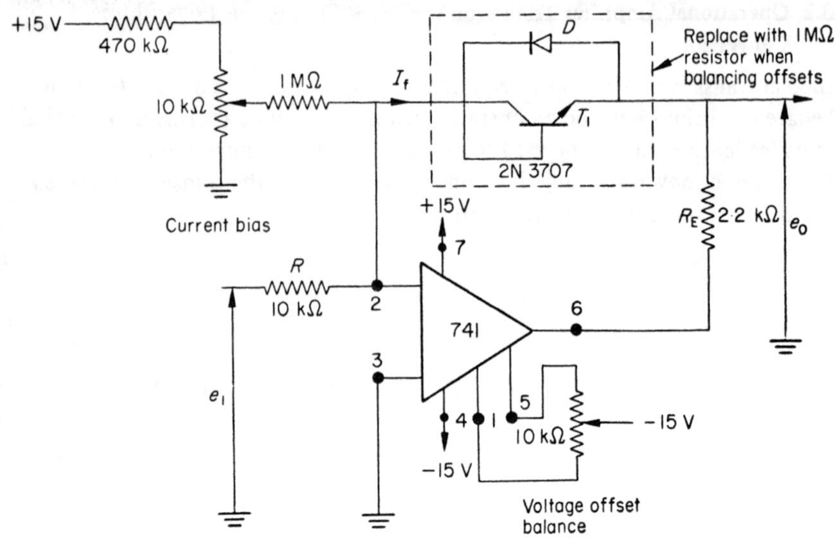

Fig. 3.3   A simple converter

transistor remains large we can neglect the base current of the transistor in comparison with its collecter current and equate its collecter current to the feedback current I_f. The output voltage of the operational amplifier supplies the emitter base voltage of the transistor and using equation 3.1 we may write

$$e_0 = V_{EB} = -E_0 \log_{10} \frac{I_c}{I_0} \tag{3.2}$$

where

$$I_c = I_f = \frac{e_i}{R}$$

Several practical points about the circuit are worth noting. The output voltage from the circuit is taken from the emitter of the logging transistor and not from the amplifier output terminal. The resistor R_E connected in series with the transistor reduces the effective loading on the amplifier output at the higher values of feedback current. The diode D connected in parallel with the logging transistor is used to protect the transistor against an excessive inverse voltage which would otherwise arise if an input voltage of wrong polarity were inadvertently applied to the circuit. The diode does not give accurate logging of negative input signals, the circuit like all log converters, is suitable only for single polarity input signals. If it is required to log negative input signals connections on both diode and transistor should be reversed.

The response equation for the circuit (equation 3.2), should be checked by applying a range of input voltages and measuring and recording input and output signals. If the widest logging range possible with the circuit is to be realised it is necessary separately to balance both the input voltage offset and the bias current of the amplifier. It is not practicable to balance offsets with the logging element in the circuit; in making offset adjustments the logging transistor together with its protective diode is removed from the circuit and replaced by a large value resistor, say 1 MΩ. Input offset voltage is balanced first, theoretically one should be able to accomplish this by shorting pin 2 to earth and adjusting the 10 kΩ offset voltage potentiometer to make the amplifier output voltage zero. In practice it will normally be found impossible to set the output voltage to zero under these circuit conditions. It is suggested that when making this adjustment the current bias be removed and pin 2 be connected to earth through a 100Ω resistor; with amplifier bias current typically of order 0.1 μA the input offset voltage remaining after the adjustment will be of order 10 μV. Once the input

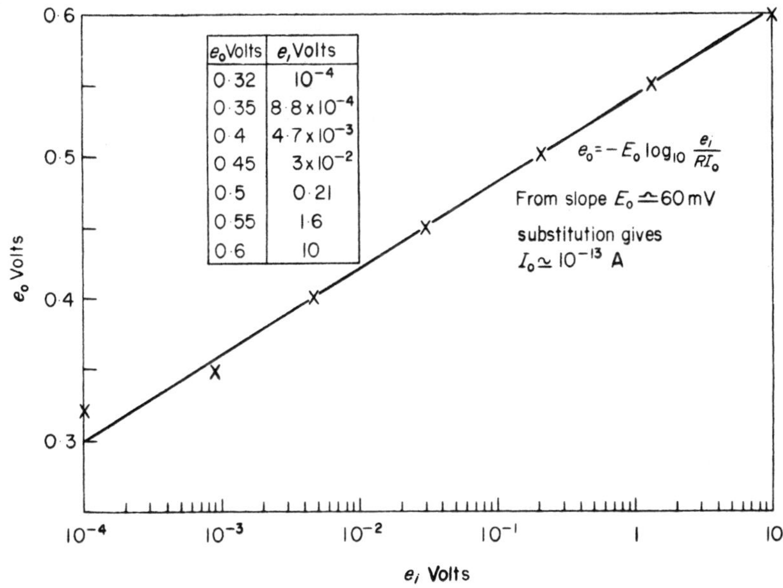

Fig. 3.4   Results obtained with a simple log converter

offset voltage adjustment has been made the connection between pin 2 and earth is removed. The input point to the circuit is connected to earth and the bias current potentiometer is adjusted so that the amplifier output is again zero. This completes the offset biasing procedure and the logging transistor with its protective diode should now be connected back into the circuit.

When investigating the logging range of the circuit, input voltages in the range say 0.1 mV to 10 mV will be found suitable. A simple resistive potential divider may be used at the input in order to measure and apply the smaller input voltages. A typical set of results are tabulated and presented graphically in figure 3.4 and are used to deduce the values of the constants E_0 and I_0 of equation 3.2.

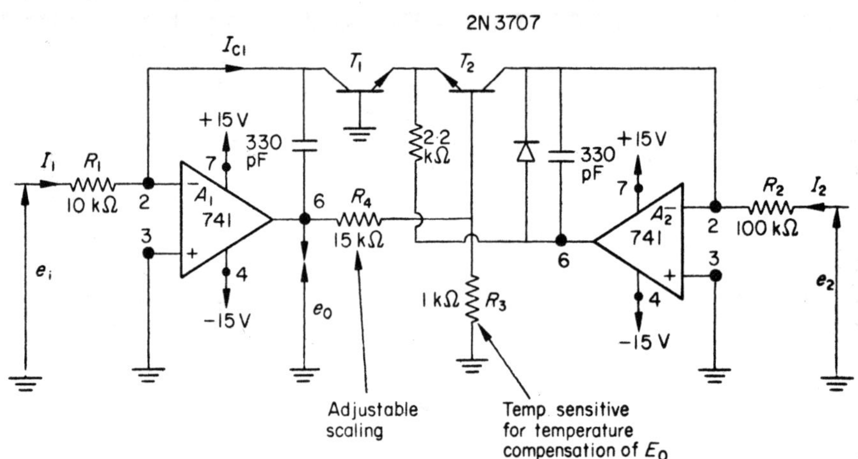

Fig. 3.5   Temperature compensated log converter

3.2.2 *A Temperature Compensated Log Converter*

A method of reducing the effects of the temperature dependence of transistor logging elements is illustrated by the practical circuit given in figure 3.5. The circuit uses two operational amplifiers and two logging transistors; its operation is now outlined. The output voltage of amplifier A_1, attenuated by the resistive divider R_3 R_4, provides the emitter base differential voltage between the transistors T_1 and T_2 and we may write

$$e_0 \frac{R_3}{R_3 + R_4} = V_{EB1} - V_{EB2} \qquad (3.3)$$

The value of V_{EB2} is controlled by negative feedback round the amplifier A_2. The feedback forces V_{EB2} to take on that value which will cause the collecter current, $I_{c2} = I_2$, to flow in transistor T_2. Negative feedback round amplifier A_1 similarly forces V_{EB1} to take on that value which will cause the collector current $I_{c1} = I_1$ to flow in transistor T_1. Substituting V_{EB} values obtained from equation 3.1 into equation 3.3 and rearrangement gives

$$e_0 = -\frac{R_3 + R_4}{R_3} E_0 \log_{10} \frac{I_{c1}}{I_{c2}} \frac{I_{02}}{I_{01}} \qquad (3.4)$$

where

$$I_{c1} = I_1 = \frac{e_1}{R_1} \quad \text{and} \quad I_{c2} = I_2 = \frac{e_2}{R_2}$$

Transistor I_0 terms have a marked temperature dependence but if the transistors in the circuit are matched the I_0 terms cancel and their temperature dependence does not cause changes in output voltage. Even if the transistors are not perfectly matched the effects of their I_0 changes are considerably reduced by the action of the circuit, for it is found that if the transistors are of the same type the ratio I_{02}/I_{01} remains fairly constant with change in temperature. The scaling factor of the converter is determined by the product of the temperature dependent E_0 terms and the ratio $(R_3 + R_4)/R_3$. E_0 has a linear temperature dependence which can be compensated by using a temperature sensitive resistor for R_3.

In order to obtain the widest possible logging range with the circuit the input offset voltage and bias current of the amplifier A_1 should be balanced using the procedure previously outlined for the simple converter. The input signal to be logged is then applied at e_1 and a fixed collector current I_{c2} set by e_2 and R_2 is passed through transistor T_2. If very small input signals are not to be used and it is required merely to take measurements in order to explore the action of the circuit it is not necessary to balance the amplifier A_1 offsets. In a practical temperature compensated log converter it is usual to return the e_2 input to the positive supply and to choose the value of R_2 to give the required value for I_{c2}. The value used for I_{c2} determines the value of I_1 and hence e_1 required for zero crossing of the output of amplifier A_1.

It is suggested that the output voltage of the circuit be measured for a range of values of e_1. Readings should be taken for several fixed values of the reference current I_2. Results are conveniently displayed by plotting the output voltage against the log of the input voltage (or input current), the slope of the graphs is equal to $E_0 (R_3 + R_4)/R_3$. Values of R_3 and R_4 are normally chosen so as to give an output voltage change of 1 V for each decade change of input current. Some typical experimental readings are shown in figure 3.6; two settings of e_2 were used, 1 volt and 10 volts, corresponding to values of I_2 of 10^{-5} A and 10^{-4} A respectively. Note that the zero crossing of the output occurs in each case when I_{c1} is slightly less than I_2. This is because of a mismatch in transistor I_0 terms. The results indicate a value $I_{01}/I_{02} \cong 0.7$ for the two transistors in the circuit. In both sets of results the accuracy of log conversion falls off for values of the input voltage less than 10 mV. Offsets in amplifier A_1 were not balanced.

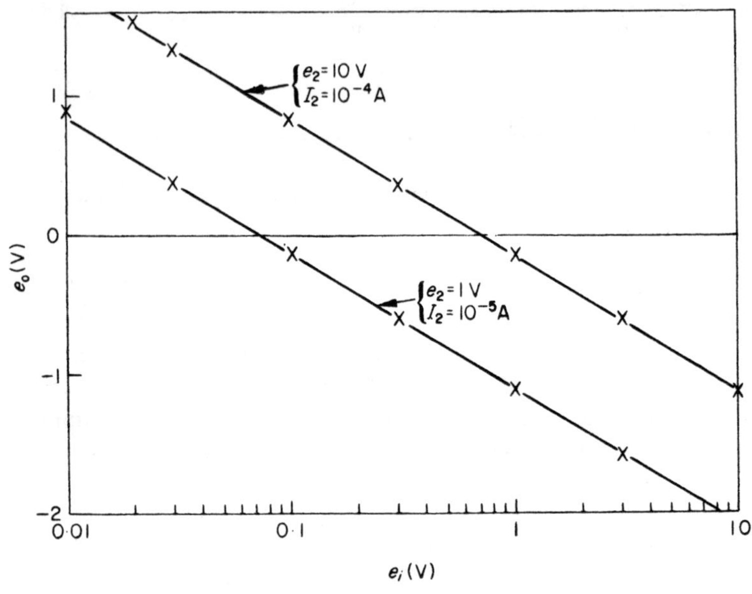

Fig. 3.6

The effect of fixing $I_{01} = I_1$ at some reference value and varying the input signal to the e_2 terminal should be tried. When used in this way the circuit gives log conversion without sign inversion but the e_2 input is not suitable for very small signals. Transistor T_2 does not give accurate logarithmic conversion for very small currents because its collecter base voltage is not maintained at zero.

Note that all operational amplifier, transistor feedback, log converters will accept only single polarity input signals. The circuit of figure 3.4 is suitable for positive input signals; if it is required to perform a logarithmic operation on a negative input signal the n-p-n transistors T_1 and T_2 should be replaced by a suitable p-n-p type (say type 2N4058).

3.3 Antilog Converters

The logarithmic characteristic of a bipolar transistor can be utilised in order to perform an antilog conversion. Simple antilog convertors, using a single transistor suffer from a marked temperature dependence but this is largely overcome by temperature compensating circuits using matched transistor pairs.

3.3.1 *A Simple Antilog Converter*

Interchanging the position of the input resistor and logging element in the simple log convertor circuit of figure 3.3 gives the circuit shown in figure 3.7 which can be used to perform an antilog conversion. The circuit is suitable for positive input voltages but a reversal of the leads to the diode connected transistor allows negative input signals to be handled. An output voltage proportional to the antilog of the input voltage is generated because negative feedback forces the input current to the circuit to flow through the resistor R and the amplifier develops an output voltage $e_0 = -I_i R$.

The input current flows as the collector current of the transistor, the applied input voltage supplying the necessary emitter base driving voltage

$$V_{EB} = -e_i, \quad I_c = I_i = -\frac{e_0}{R}$$

substitution in equation 3.1 and rearrangement gives

$$e_0 = -R I_0 \; 10^{\frac{e_i}{E_0}} \tag{3.5}$$

In order to check the behaviour of the circuit values of e_i in the range, say 300 mV to 600 mV should be applied and input and output voltages measured and recorded. Results are conveniently displayed graphically as $e_i/\log_{10} e_0$.

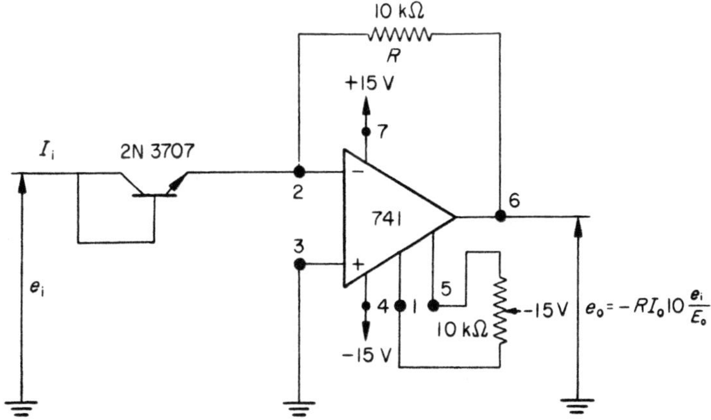

Fig. 3.7   Simple antilog converter

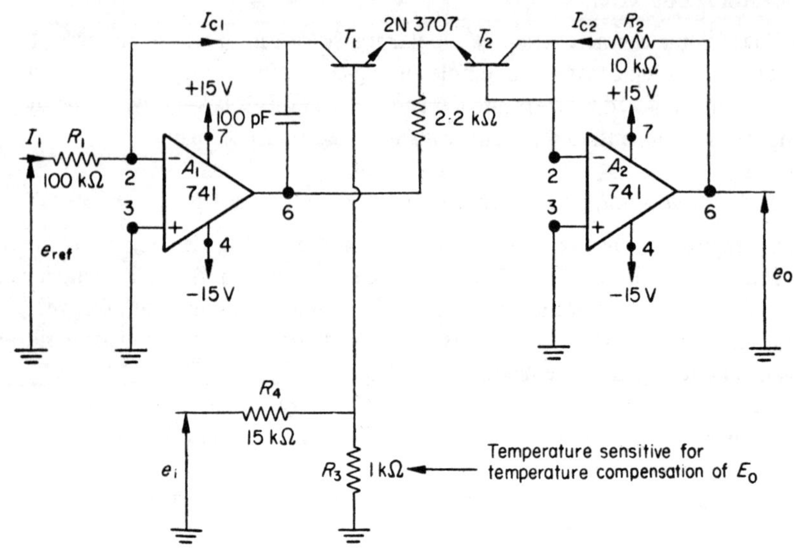

Fig. 3.8 Temperature compensated antilog converter

3.3.2 *A Temperature Compensated Antilog Converter*

A temperature compensated circuit for antilog conversion is illustrated in figure 3.8. It is basically a rearrangement of the circuitry in the temperature compensated log converter previously introduced in figure 3.5.

The input signal to the circuit, attenuated by the resistive divider R_3 R_4, provides the emitter base differential voltage between transistors T_1 and T_2 and we may write

$$e_i \frac{R_3}{R_3 + R_4} = V_{EB2} - V_{EB1} \tag{3.6}$$

Negative feedback around amplifier A_1 forces V_{EB1} to take on that value which will cause the current $I_1 = I_{c1}$ to flow as a collector current in transistor T_1. If I_1 is held constant, as a reference current, V_{EB1} is constant and V_{EB2} varies directly with the input voltage e_i. V_{EB2} determines the collector current I_{c2} of transistor T_2. Negative feedback around amplifier A_2 forces I_{c2} to flow through resistor R_2 and amplifier A_2 develops an output voltage $e_0 = I_{c2} R_2$. Substitution of V_{EB} values from equation 3.1 into equation 3.6 gives

$$e_i \frac{R_3}{R_3 + R_4} = E_0 \log_{10} \frac{I_{c1}}{I_{c2}} \frac{I_{02}}{I_{01}}$$

where

$$I_{c2} = \frac{e_0}{R_2} \quad \text{and} \quad I_{c1} = I_1 = \frac{e_{ref}}{R_1}$$

Substituting for I_{c2} and rearranging gives

$$\frac{I_{02}}{I_{01}} I_{c1} \frac{R_2}{e_0} = 10 \, e_i \frac{R_3}{R_3 + R_4} \frac{1}{E_0}$$

The linear temperature dependence of E_0 may be compensated for by using a temperature sensitive resistor for R_3. Values of R_3 and R_4 are normally chosen so as to make $R_3/(R_3 + R_4)$ $1/E_0$ unity and assuming they have been so chosen the equation becomes

$$e_0 = I_{c1} \frac{I_{02}}{I_{01}} R_2 \, 10^{-ei} \tag{3.7}$$

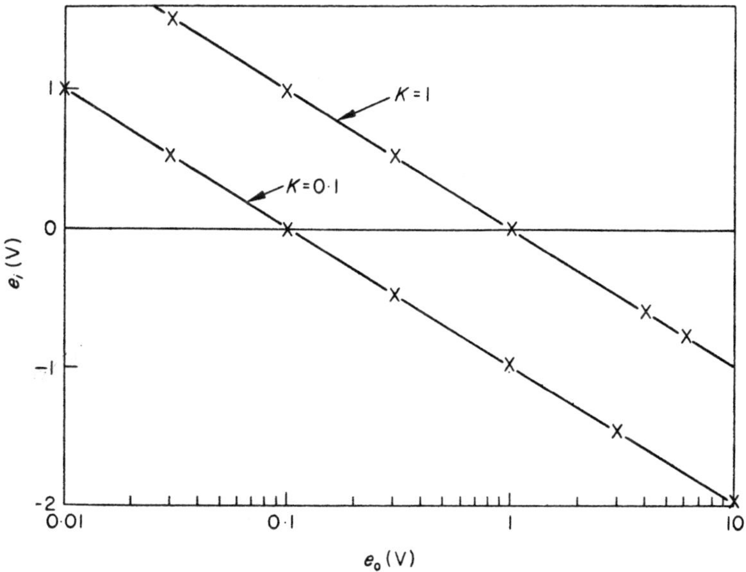

Fig. 3.9   Results obtained with temperature compensated antilog converter

The multiplying factor, $K = I_{02}/I_{01}$ $(I_{c1} R_2)$ is made equal to a desired constant by choice of e_{ref}, R_1, and R_2, $(I_{c1} = e_{ref}/R_1)$. K must be chosen so that the output voltage capability of the amplifier is not exceeded when the input signal has its maximum expected value.

The following practical setting up procedure for the circuit is suggested, it allows for transistor mismatch and avoids the use of close tolerance resistors. Set e_i to zero and adjust e_{ref} or R_1 to make the output voltage of amplifier A_2 exactly K volts. Apply an input signal of minus one volt and trim the value of resistor R_4 to make the output of amplifier A_2 exactly 10 volts. Typical experimental results obtained with the circuit after carrying out this procedure are presented graphically in figure 3.9. The two sets of results are for $K = 1$ and $K = 0.1$; no offset balance was employed, balancing amplifier A_2 offsets may be expected to extend the range of the circuit.

3.4 Log Circuits for Multiplication, Division and the Generation of Powers

Operational amplifier log and antilog converters may be combined in order to generate many non-linear functions. The circuits are connected together in such a way that they perform the operations normally involved in logarithmic computations. Examples of such computations are described by the equations

antilog $(n \log x) = x^n$
antilog $(\log x + \log y) = x y$
antilog $(\log x - \log y) = x/y$

3.4.1 *Power Generator*

In order to generate an output signal proportional to the nth power of an input signal a log converter is used to generate the log term. A resistive divider is used to multiply the log term by a constant n and an antilog converter is then used to form the output signal. An experimental circuit for investigating the performance of a power generator is given in figure 3.10. The circuit consists essentially of a combination of the temperature compensated log converter and temperature compensated antilog converter previously described. A brief analysis of its action is now given.

Referring to figure 3.10

$$V_{EB4} = V_{EB3} + e_{01} \frac{R_8}{R_7 + R_8} \tag{3.8}$$

Now by equation 3.1

$$V_{EB3} = -E_0 \log_{10} \frac{I_{c3}}{I_{03}}$$

and

$$V_{EB4} = -E_0 \log_{10} \frac{I_{c4}}{I_{04}}$$

The output of amplifier A_1 is given by equation 3.4

$$e_{01} = -\frac{R_5 + R_6}{R_6} E_0 \log_{10} \frac{I_{c1}}{I_{c2}} \frac{I_{02}}{I_{01}}$$

Substituting values in equation 3.8 gives

$$E_0 \log_{10} \frac{I_{c3}}{I_{03}} + \frac{R_8}{R_7 + R_8} \frac{R_5 + R_6}{R_6} E_0 \log_{10} \frac{I_{c1}}{I_{c2}} \frac{I_{02}}{I_{01}} = E_0 \log_{10} \frac{I_{c4}}{I_{04}}$$

It is assumed that the transistors are all at the same temperature, which makes the E_0 terms equal and a rearrangement of the equation gives

$$\log_{10} \frac{I_{c4}}{I_{c3}} \frac{I_{03}}{I_{04}} = n \log \frac{I_{c1}}{I_{c2}} \frac{I_{02}}{I_{01}}$$

where

$$n = \frac{R_8}{R_6} \frac{R_5 + R_6}{R_7 + R_8}$$

If it is further assumed that the transistors are matched the I_0 terms cancel and the equation becomes

$$\frac{I_{c4}}{I_{c3}} = \left[\frac{I_{c1}}{I_{c2}} \right]^n$$

Now

$$I_{c4} = \frac{e_0}{R_4}, \quad I_{c3} = \frac{e_3}{R_3}, \quad I_{c2} = \frac{e_2}{R_2} \quad \text{and} \quad I_{c1} = \frac{e_1}{R_1}$$

thus

$$e_0 = \frac{e_3}{R_3} R_4 \left[\frac{R_2}{e_2 R_1} \right]^n e_1^n \tag{3.9}$$

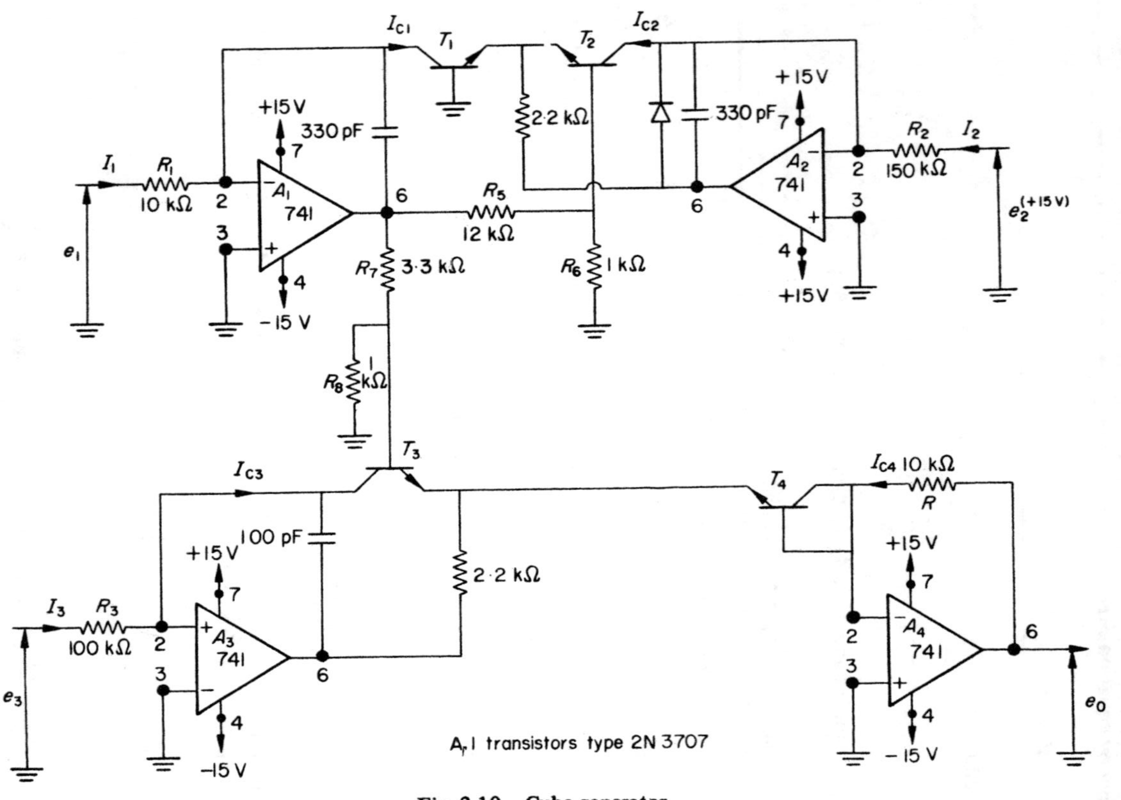

Fig. 3.10   Cube generator

The component values in figure 3.10 are chosen to make the scaling factor of the generator unity and the power $n = 3$. It is suggested that the voltage e_3 is made variable to allow for a mismatch in the I_0 terms. In setting up the circuit for operation an input signal of exactly one volt is applied and e_3 is adjusted to make the output signal exactly one volt. Experimental results obtained with the circuit are illustrated graphically in figure 3.11. In one case the component values indicated in the circuit diagram were used to make $n = 3$, in the other case components were changed to make $n = \frac{1}{2}$. Values $R_6 = R_8 = 1\ k\Omega$, $R_5 = 5.6\ k\Omega$ and $R_7 = 12\ k\Omega$ were used.

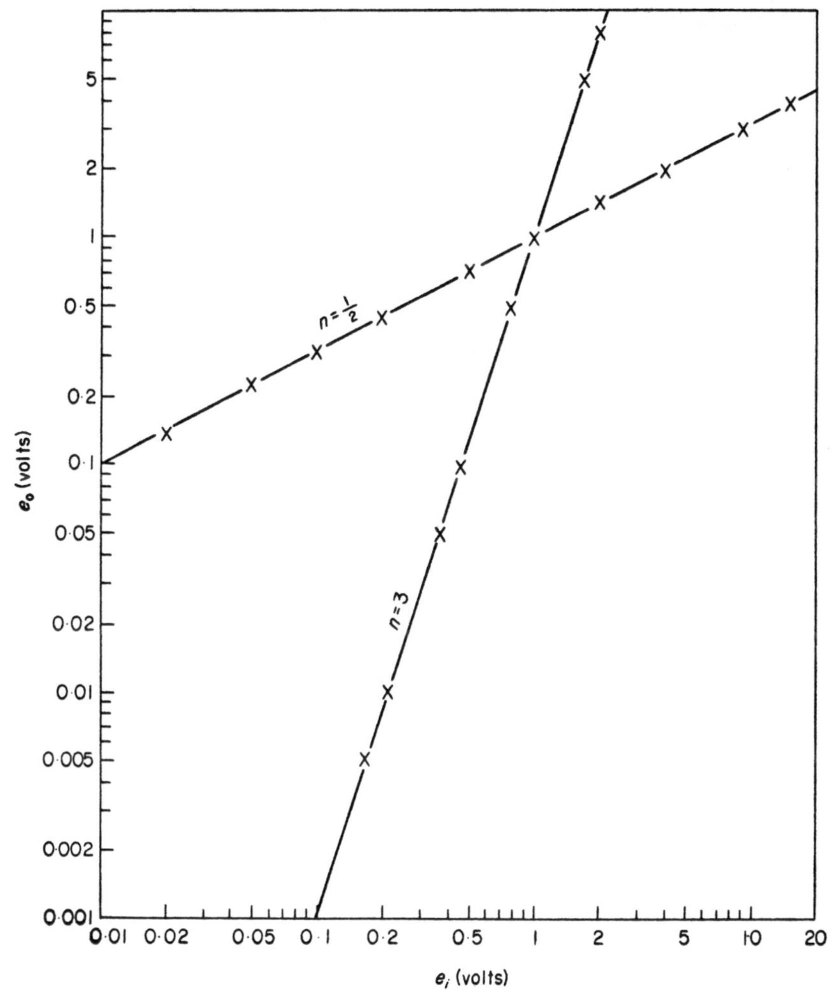

Fig. 3.11   Experimental results obtained with power generator

The lines show the calculated functions, $e_0 = e_1^3$ and $e_0 = e_1^{1/2}$ respectively and the plotted points indicate experimentally obtained data; 5 per cent tolerance resistors were used to set powers. If greater conversion accuracy is required resistor values should be selected in order to set precisely a required power. Accuracy at low signal levels can be improved by balancing the offsets of amplifiers A_1 and A_4.

3.4.2 *Multiplier/Divider*

Examination of equation 3.9 shows that if the power n is made unity the response of the power generator circuit of figure 3.9 becomes

$$e_0 = \frac{R_4}{R_3} \frac{R_2}{R_1} \frac{e_3\, e_1}{e_2} \tag{3.10}$$

The power n may be set to unity by an appropriate choice of log scaling resistors or the scaling resistors may be omitted altogether as shown in the multiplier/divider circuit of figure 3.12. In this circuit the log output at the base of transistor T_2 is connected directly to the antilog circuit at the base of transistor T_3. The circuit response is described by equation 3.10 and the circuit can be used for single quadrant multiplication or division. All signals must be positive; if a circuit which will handle negative signals is required the n-p-n transistors in the circuit should be replaced with p-n-p types.

In a multiplier application of the circuit the signals to be multiplied are applied to inputs e_1 and e_3. Scaling, according to equation 3.10, is determined by the values of the resistors R_1, R_2, R_3 and R_4 and by the signal e_2. In practice, because of mismatch in transistor I_0 terms, it is usually necessary to make one of the scaling parameters adjustable. A convenient practical procedure is to fix the resistor values, apply measured values of e_1 and e_3 and adjust the value of e_2 to give the output product multiplied by the desired scaling factor. When using the circuit for division the input signals are applied to $e_1\, e_2$ and e_3 can then be adjusted to fix the scaling factor.

3.5 Log Circuits – further Practical Considerations

The logarithmic circuits presented in this chapter have been connected up by the author, in bread-board form, for the purpose of evaluation. The circuits work but would benefit from minor modifications if they are to be developed as practical applications.

The closed loop stability of logarithmic circuits requires particular attention[1]. Stray input capacitance and load capacitance, which is governed by the nature of the actual physical circuit arrangements, has an effect on the values of the

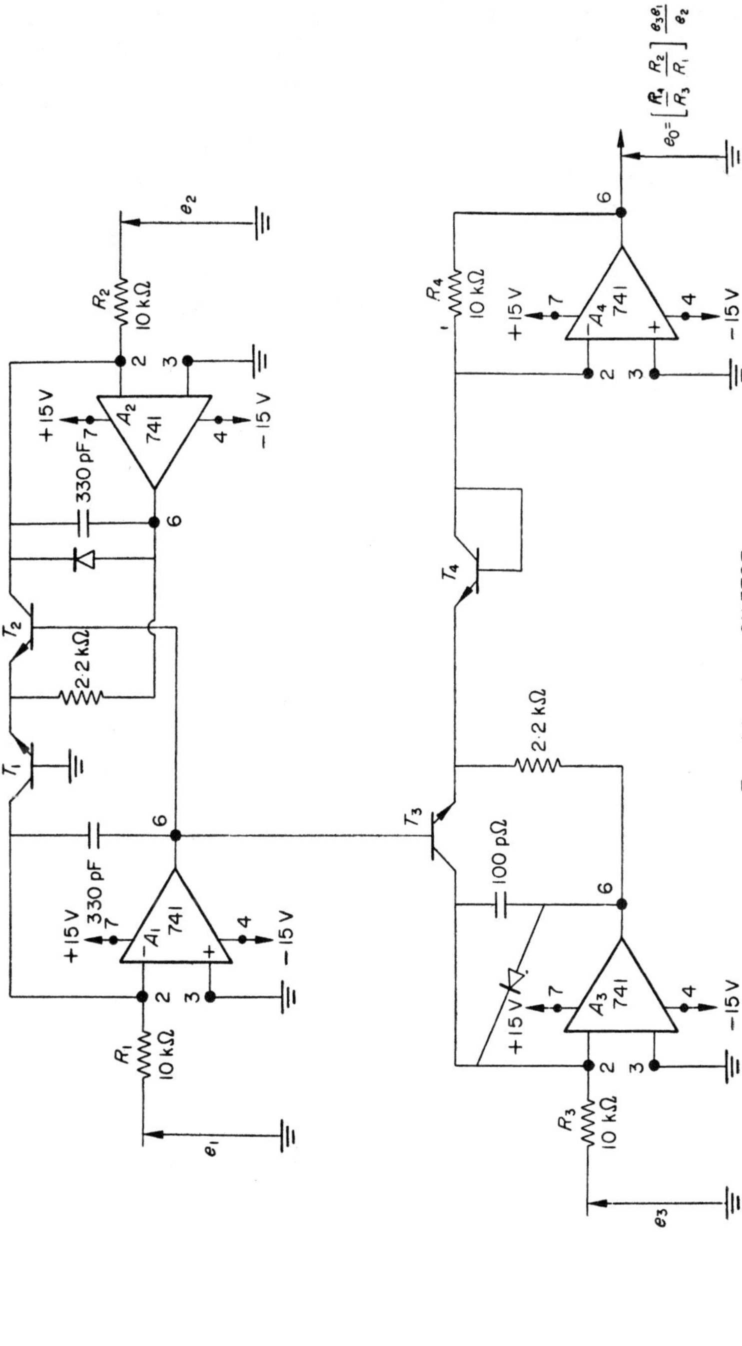

Transistors type 2N 3707

Fig. 3.12 Multiplier/Divider

frequency compensating feedback capacitors which are used in the circuits to ensure closed loop stability. The reader may find it necessary to use alternative values for frequency compensating capacitors; closed loop stability should always be experimentally checked, an oscilloscope can be used to verify that no circuit oscillations are present. Values of frequency compensating capacitors control the 'speed' of log circuits, that is, they determine the rate at which the output voltage can change, the larger the values of the capacitors the slower are the circuits. Circuits are slowest when input currents have their minimum value.

The effectiveness of the temperature compensation circuit techniques used in log converters is determined by the matching of the logging transistors; temperature differentials between the transistors should be avoided. The effect of a temperature differential is easily demonstrated in the experimental circuits by simply hand warming one of the logging transistors. The use of dual matched transistors should be considered in practical log circuits and the use of alternative lower drift amplifier types may be expected to extend the dynamic range of the circuits.

3.6 Further Applications of Log Circuits

Various combinations of log and antilog circuits can be used together with operational amplifier circuits to perform a wide variety of functional operations; single log converters are very useful in obtaining a wide dynamic range in signal processing systems. In linear systems there is a marked loss of accuracy when the input signal is small compared with full scale but in the case of log amplifiers accuracy is a percentage of signal, rather than a percentage of full scale, over most of the dynamic range. Accuracy over a wide dynamic range is also a feature of the logarithmic multiplier/divider/power generation circuits. Four-quadrant linear multipliers[5], can be used for multiplication, division, squaring and square rooting, but logarithmic circuits give greater accuracy for wide range signals.

3.6.1 *Log Divider used for Transistor Current Gain Measurement*

A comparatively simple application of the log multiplier/divider circuit is shown in figure 3.13; a practical arrangement is outlined for the measurement of the current gain of a transistor over a range of operating currents. The circuit in figure 3.13 is used in conjunction with the multiplier/divider circuit of figure 3.10. The collector current of the transistor supplies the input current to amplifier A_1 and the base current of the transistor provides the input current to amplifier A_2; resistors R_1 and R_2 are not required and are omitted from the circuit of figure 3.10. The output of the divider circuit is proportional to the current gain of the transistor, I_c/I_B; scaling may be set by adjusting e_3. The operating current of the transistor is varied by changing the value of the positive voltage supply which is connected to the emitter of the transistor by means of the resistor R.

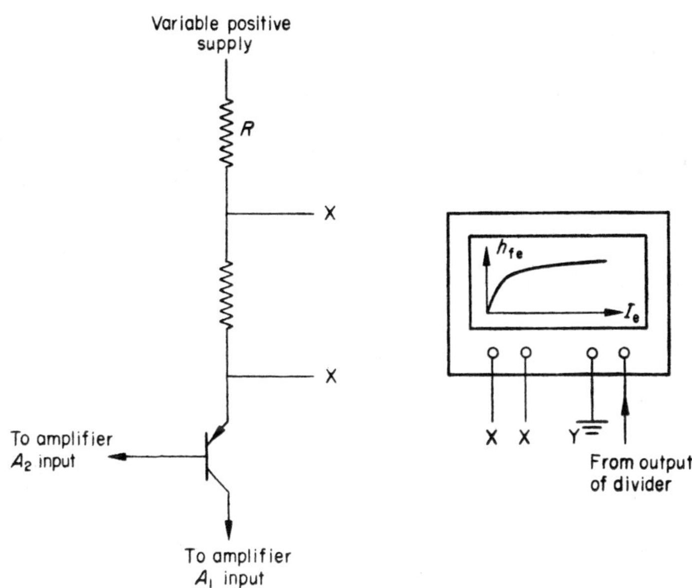

Fig. 3.13   Measurement of h_{fe} variation with divider

3.6.2 *Multifunction Logarithmic Circuit Modules*

Ready built circuit modules are available which employ log/antilog techniques similar to those discussed in sections 3.4.1 and 3.4.2. The designer with a signal processing application requiring multiplication, division or exponentiating of single polarity signals over a wide dynamic range may find it worthwhile to use such modules rather than to develop his own computation circuits. The relative costs of buying or building are much influenced by the cost of the engineers designing and building time.

An example of a circuit module which employs log/antilog techniques is Analog Devices type AD 433[34]. The performance of this circuit module is characterised by the function

$$V_0 = \frac{10}{V_R} \quad V_Y \left[\frac{V_Z}{V_X}\right]^n \tag{2.11}$$

Note the similarities between this equation and equation 3.9. Two external resistors connected to the AD 433 circuit module are used to set a value for the exponent n; values can be set in the range $1/5 \leqslant n \leqslant 5$.

References

1. G.B. Clayton, *Operational Amplifiers*. Butterworths (1971).

2. W. Bolase and E.I. David, Design of temperature compensated log circuits employing transistors and operational amplifiers. *Application Report Analog Devices.*

3. F. Pouliot and L. Counts. $Y \ [\frac{z}{x}]^m$ at low cost, *Analog Dialogue* Vol. 6, No. 2.

4. D. Sheingold, Trigonometric Operations with the 433. *Analog Dialogue* Vol. 6, No. 3.

5. G. B. Clayton, *Linear IC Applications Handbook*. TAB BOOKS, (1977).

Exercises 3

3.1   If reference voltages are $\pm$ 10 volts and diodes behave ideally what values of the resistors R_a and R_b should be used in the circuit of figure 3.1 in order that the circuit shall provide an input output relationship shown in figure E.3.1 [R_f = 100 kΩ]

3.2   A simple log converter is shown in figure E.3.2. The circuit is required to give an output voltage change of 3 volts when the input changes from 0.1 V to 100 V. What values should be used for the resistors R_1 and R_2? Discuss the limitations of the simple circuit.

3.3   In the temperature compensated log converter (figure 3.5) R_1 = 10 kΩ, R_2 = 200 kΩ, R_3 = 1 kΩ. What should the value of R_4 be for a scaling factor of 1 V per decade change of the input signal e_1. If the transistor I_0 terms are in the ratio I_{01}/I_{02} = 0.5, what must be the value of e_2 in order that the circuit shall give zero output voltage when e_1 = 0.1 volts.

3.4   Using the circuit of figure 3.10 it is required to produce an output signal of the form $e_0 = e_1$ 3/2. The following component values are used in the circuit; R_6 = 1 kΩ, R_8 = 1 kΩ, R_5 = 15 kΩ, R_1 = 10 kΩ, R_2 = 100 kΩ, R_3 = 100 kΩ, R_4 = 10 kΩ, e_2 = 10 volts.

   Transistor I_0 terms are in the ratio I_{01}/I_{02} = 0.5, I_{03}/I_{04} = 0.8. Find the value required for the resistor R_7 and the reference voltage e_3.

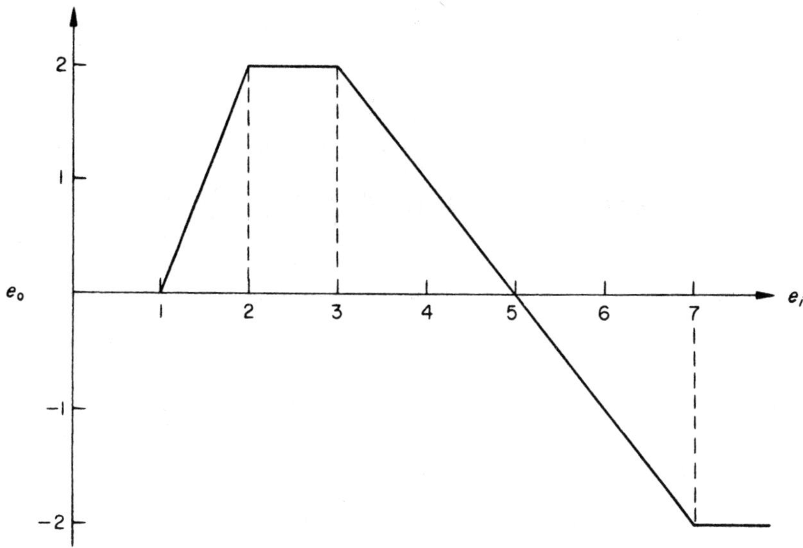

Fig. E.3.1 Input/output relationship to be generated in question 3.1

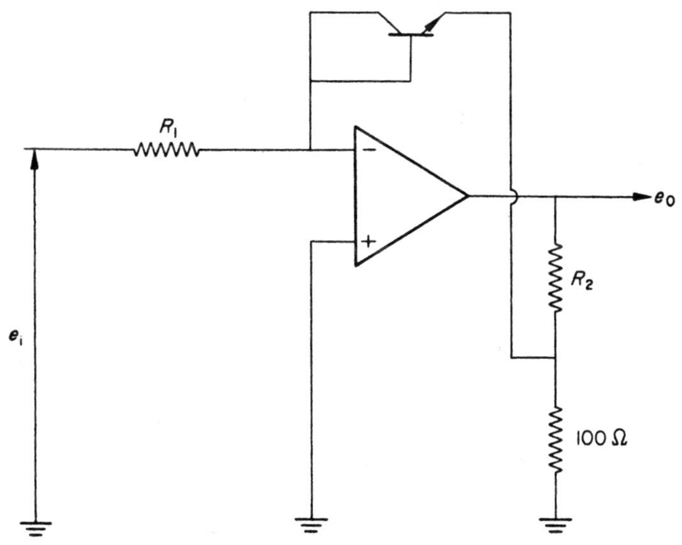

Fig. E.3.2 Simple log converter of question 3.2

4. Some Signal Processing and Measurement Applications

4.1 Precise Rectification with an Operational Amplifier/Diode Combination

Semiconductor diodes show pronounced non-linearity at low forward voltages. In the case of a silicon diode no appreciable conduction takes place through the diode until the forward voltage across it exceeds 0.5 V. As a result of this non-linearity diodes give rise to appreciable errors when they are used to rectify small signals in a conventional rectification circuit.

The circuit in figure 4.1 shows a way in which an operational amplifier can be used to overcome the effects of diode non-linearities. In this circuit the diode is connected in the feedback path of the amplifier, the initial forward voltage required to cause diode conduction is supplied by the amplifier output voltage. The input signal required by the amplifier is extremely small, for if the diode is non-conducting there is effectively no feedback and the amplifier has its full

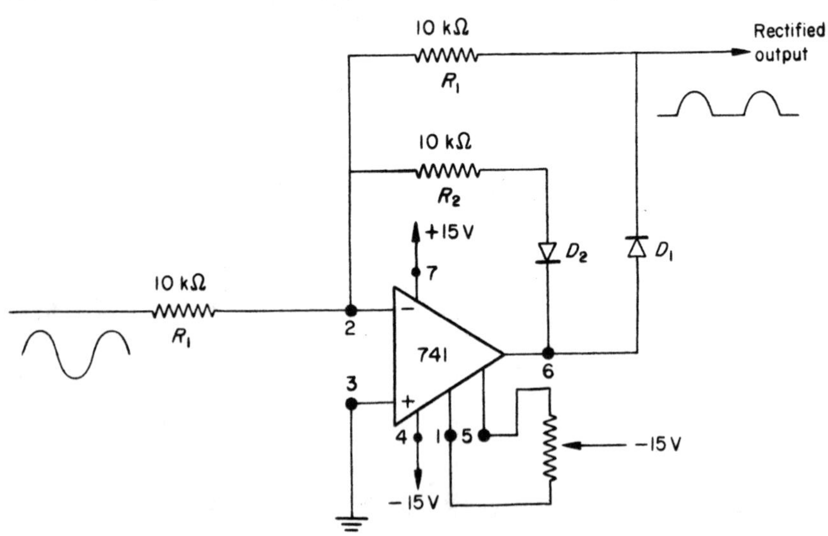

Fig. 4.1   Precise rectification

open loop gain. Once the diode has been brought into conduction the effect of its non-linearities on the rectified output is effectively divided by the loop gain.

Typical waveforms obtained with the circuit of figure 4.1 are shown in figure 4.2, the upper trace shows a sinusoidal input signal, the middle trace is the amplifier output signal and the lower trace is the rectified output signal. The amplitude of the input signal is less than 0.2 V. A simple diode rectifier would give no measurable output with an input signal of this magnitude. The symmetrical amplifier output waveform in figure 4.2 was obtained by adjustment of the offset balance potentiometer.

The second feedback path in the circuit, through diode D_2, is included to maintain the virtual earth at the amplifier summing point on the input half cycles for which D_1 is reverse biased. Gain as well as rectification is given by the circuit if the resistor ratio R_2/R_1 is made greater than unity.

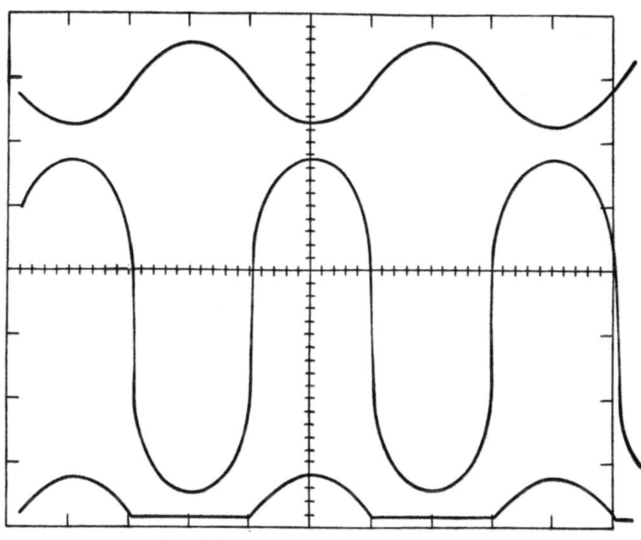

Fig. 4.2   Precise rectifier waveforms

4.1.1 *A Precise Rectifier used as a Millivoltmeter*

An operational amplifier, a diode bridge and a d.c. microammeter can be used as the basis for a precise a.c. millivoltmeter; the circuit shown in figure 4.3 is useful for investigating the action and limitations of such an arrangement. In figure 4.3 the a.c. input signal is applied to the non-phase inverting input terminal of the amplifier, feedback through the diode bridge forces the voltage across the resistor R in the circuit to follow the input signal. A current e_i/R must pass through the bridge. This current, rectified by the bridge, passes through the d.c. meter and the meter indicates the average value of the full wave rectified current.

If e_i represents the RMS value of the sinusoidal input signal the current through the d.c. meter is determined by the relationship

$$I_{av} = \frac{2\sqrt{2}}{\pi} \; \frac{e_i}{R} \tag{4.1}$$

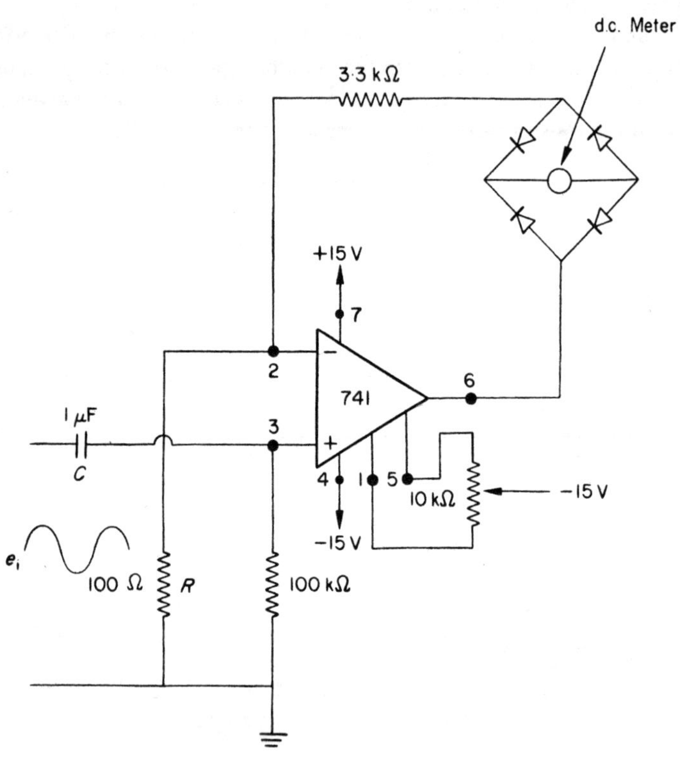

Fig. 4.3   A.C. Millivoltmeter

In investigating the action of the circuit it is convenient to measure the d.c. current with a multirange meter. The meter is initially set to one of its less sensitive ranges, the input to capacitor C is earthed and the offset balance potentiometer is adjusted for minimum current through the meter; the sensitivity of the meter is increased whilst making the adjustment. An a.c. input signal, with a frequency of say 120 Hz, is now applied. Its amplitude is increased in steps and the reading of the d.c. meter is recorded for each value of the input signal. Suitable values for the input signal are, 2 mV, 4 mV, 6 mV, 8 mV, 10 mV, (RMS values), and so on through increasing decades. The linearity of the system can now be assessed; in making the assessment do not forget possible errors or non-linearities in the test instruments. A pronounced non-linearity occurs when the amplifier output saturates. The amplifier output signal should be monitored with an oscilloscope in order to detect the onset of saturation.

The frequency response of the system can be investigated by applying an input signal of fixed amplitude, (say 50 mV), and increasing the frequency until the reading of the d.c. meter begins to fall. The reason for the fall in response at the higher frequencies will be obvious if the output waveform of the amplifier is monitored. Amplifier output waveforms for frequencies of 50 Hz and 5 kHz are shown in figure 4.4. The steps in the 50 Hz waveform occur as the amplifier output voltage takes up the diode forward voltage drops. When the frequency is increased to 5 kHz the steps in the waveform occupy an appreciable fraction of the waveform period and the average value of the rectified current through the meter falls. Finite step times arise because of amplifier slewing rate limitations, this amplifier performance parameter thus determines the upper limit of the frequency response of the system.

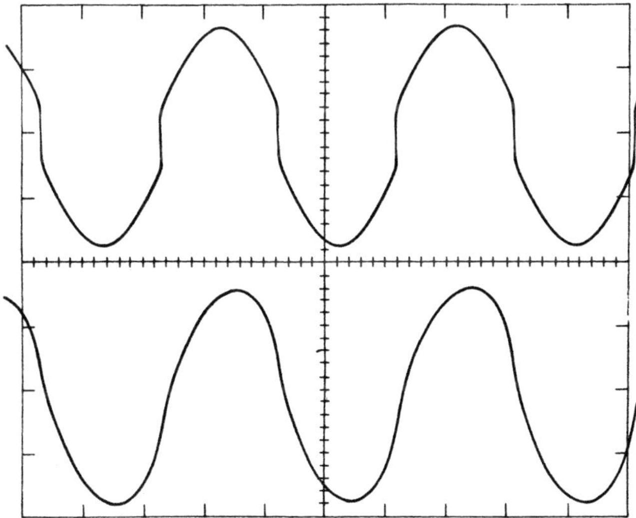

Fig. 4.4   Amplifier output waveforms of millivoltmeter circuit

4.2 Phase Sensitive Detection

A phase sensitive detector, PSD, is a system which produces a d.c. output signal in response to an a.c. input signal whose frequency is equal to that of an a.c. reference signal. The d.c. output given by a phase sensitive detector is proportional to the amplitude of the input signal and the cosine of its phase angle is relative to that of the reference signal. PSD's are used as synchronous rectifiers and in chopper d.c. amplifiers they are also used for the accurate measurement of small a.c. signals that are obscured by noise[1].

A PSD acts essentially as a multiplier giving an output signal which is proportional to the product of the input signal and the reference signal. It is instructive to analyse this multiplication process. We consider a sinusoidal signal, $e_s = E_s \sin(\omega_s t + \phi)$ multiplied by a square wave reference signal; the square wave is assumed to be symmetrical about zero and to have unit amplitude and unity mark-space ratio.

The Fourier series for a unit amplitude square wave is

$$v_s = \frac{4}{\pi} \left(\sin \omega_r t + \frac{1}{3} \sin 3 \omega_r t + \frac{1}{5} \sin 5 \omega_r t \right) \qquad (4.2)$$

and the product of the input signal and reference signal is

$$e_s v_r = \frac{4}{\pi} E_s \left[\sin(\omega_s t + \phi) \sin \omega_r t + \sin(\omega_s t + \phi) \frac{1}{3} \sin 3 \omega_r t + ... \right]$$

$$= \frac{2}{\pi} E_s \left[\cos[(\omega_s - \omega_r) t + \phi] - \cos[(\omega_s + \omega_r)]t + \phi] \right.$$

$$+ \frac{1}{3} \cos[(\omega_s - 3 \omega_r) t + \phi] - \frac{1}{3} \cos[(\omega_s + 3 \omega_r) t + \phi]$$

$$\left. + \frac{1}{5} \dots \text{etc.} \right] \qquad (4.3)$$

Examination of equation 4.3 shows that the product gives rise to d.c. terms for signal frequencies $\omega_s = \omega_r$; $3 \omega_r$; $5 \omega_r$ etc.

When $\omega_s = \omega_r$ the d.c. term is $\frac{2}{\pi} E_s \cos \phi$

when $\omega_s = 3 \omega_r$ the d.c. term is $\frac{2}{\pi} E_s \frac{1}{3} \cos \phi$ etc.

In a PSD the multiplication process is followed by low pass filtering which can be performed by a simple *RC* filter which will attenuate all a.c. components of the product. The effective bandwidth of such a system is thus set at 2/*T* radians/second where $T = CR$, the time constant of the filter. The noise rejection properties inherent in a phase sensitive detection system arise in large part from the extremely narrow bandwidth which is so readily obtained by selection of filter time constant. A PSD acts somewhat like a rectifier that is tuned to the reference frequency but its noise rejection properties are superior to a conventional filter rectifier combination. Noise passed by a conventional filter circuit gives rise to a d.c. component in the normal rectification process, but in a PSD unwanted signals, like random noise, produce fluctuations about the d.c. level given by the wanted signal at the reference frequency. It would be quite impractical to obtain the narrow bandwidth of a PSD with a conventional filter circuit.

4.2.1 *A Phase Sensitive Detector Design*

Many PSD designs of varying degrees of complexity have been described in the literature[1,2,3,4]. In general it is convenient if the signat input, the reference input and the output from the detector share a common earth line, a condition which is not always fulfilled in all PSD designs. An operational amplifier form of PSD which fulfils the above conditions is illustrated in figure 4.5.

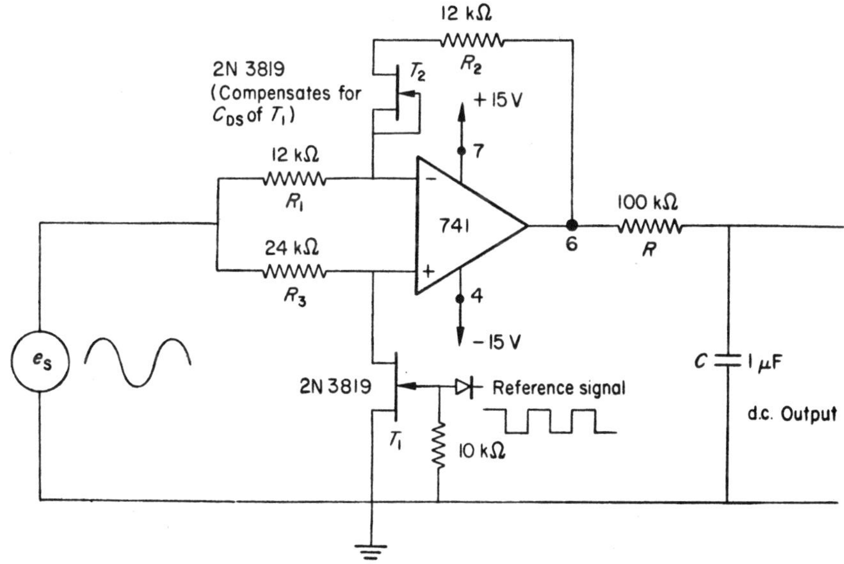

Fig. 4.5   An operational amplifier used as a phase sensitive detector

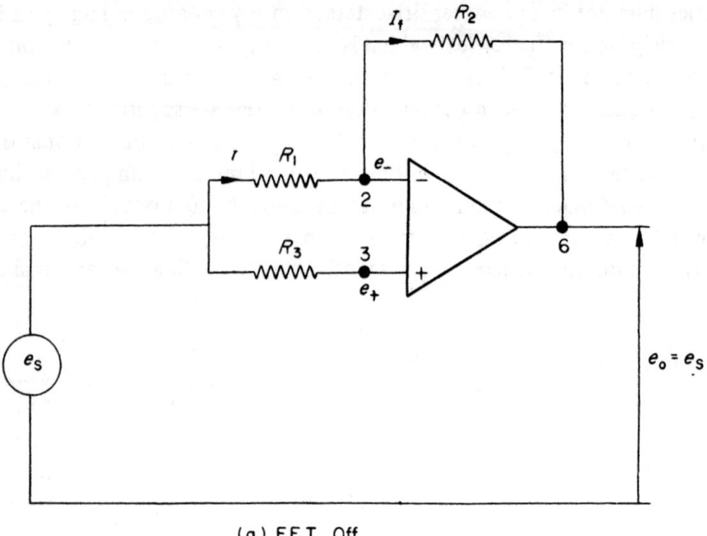

(a) F.E.T. Off

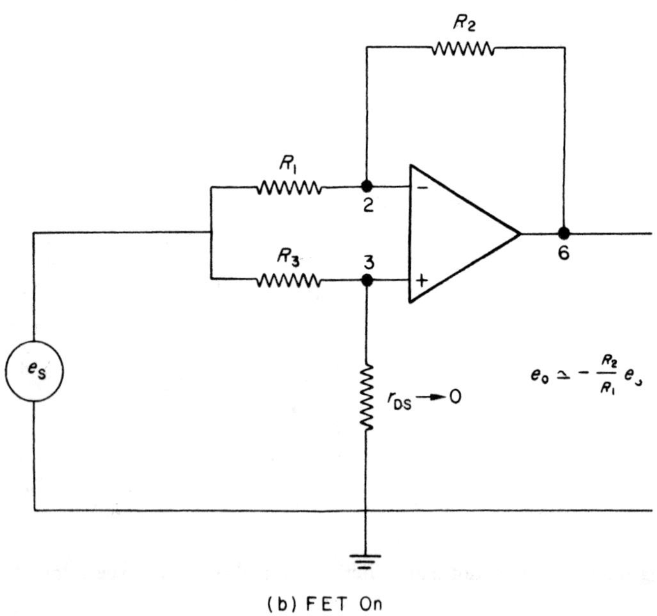

(b) FET On

Fig. 4.6

In the circuit of figure 4.5 the closed loop gain of the operational amplifier is switched between plus and minus unity by a square wave reference signal which is applied to the gate terminal of the FET in the circuit. The circuit is very convenient for illustrating phase sensitive detector operation. Conditions with the FET biased OFF and ON are shown in figures 4.6(a) and 4.6(b) respectively. Referring to figure 4.6(a) and assuming that the operational amplifier behaves ideally we have

$$e_- = e_+ = e_s$$

thus $I_i = I_f = 0$ and e_0 must be equal to e_s. The closed loop gain is independent of resistor values and has a value of plus unity. In figure 4.6(b) r_{ds} represents the effective drain source resistance of the conducting FET. If $r_{ds} \rightarrow 0$ the non-phase inverting input terminal of the amplifier is effectively earthed and the circuit acts as a phase inverting feedback amplifier with closed loop gain $-R_2/R_1$. A non-zero value of r_{ds} allows a small fraction of the input signal to appear at the non-phase inverting input terminal of the amplifier and R_2 must be made slightly larger than R_1 for a closed loop gain of minus unity. (use T_2 with r_{ds} matched to T_1).

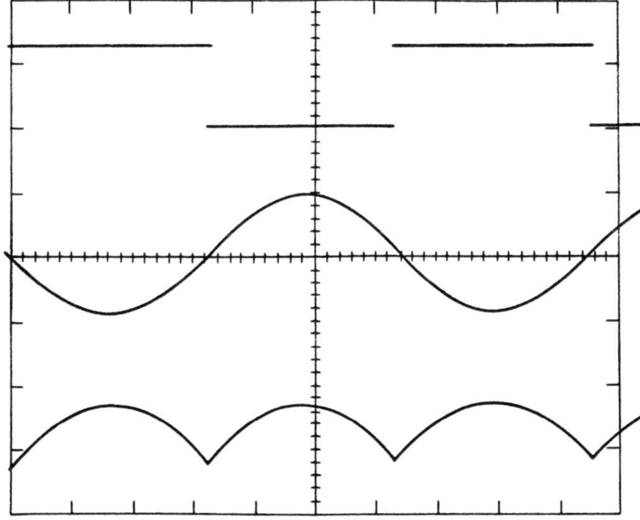

Fig. 4.7   Signal and reference in phase; upper trace 10V/division; middle and lower trace 5V/division; horizontal 2ms/division

The action of the PSD is illustrated by figures 4.7 to 4.10 in which waveforms are shown for various phase relationships between input and reference signals. In each figure the upper trace shows the reference square wave applied to the gate of the FET, the middle trace shows the sinusoidal input signal and the lower trace shows the waveform which appears at the output terminal of the amplifier. The average value of the output may be measured by a d.c. voltmeter connected across capacitor C. When making quantitative measurements with the system in order to check the validity of equation 4.3, it is convenient to derive the phase shifted reference square wave and the input signal from the same signal source. A phase shifting and squaring circuit suitable for this purpose is shown in figure 4.11.

It is instructive to observe the behaviour of the PSD for input signals at frequencies other than that of the reference signal. Two separate signal sources are required for this purpose, a reference square wave and an input sinusoid. The waveforms shown in figure 4.12 illustrate the response of the system to an input signal at a frequency equal to the third harmonic of the reference signal. It is clear that the output waveform has a non-zero average, its average value is

$$\frac{2}{\pi} \, E_s \, \frac{1}{3} \, \cos \phi$$

where in this case $\phi = 180°$.

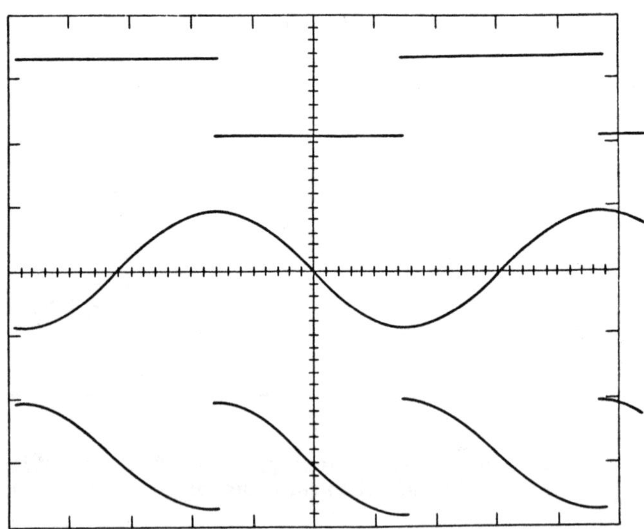

Fig. 4.8 Signal and reference phase shifted $45°$; upper trace 10V/division; middle and lower trace 5V/division; horizontal 2ms/division

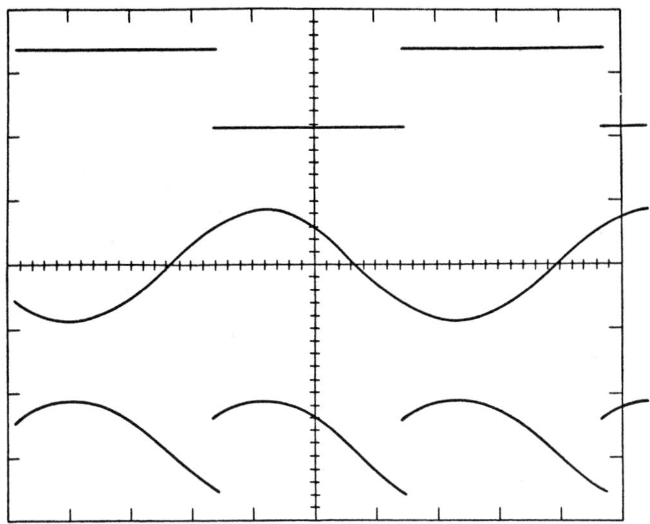

Fig. 4.9 Signal and reference shifted 90°; upper trace 10V/division; middle and lower trace 5V/division; horizontal 2ms/division

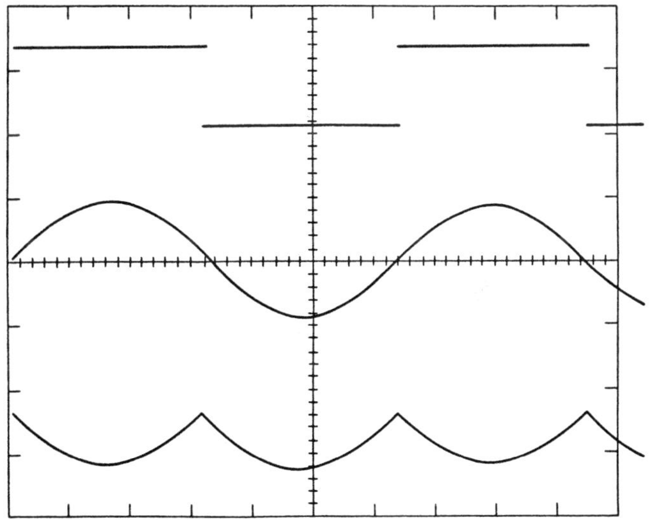

Fig. 4.10 Signal and reference phase shifted 180°; upper trace 10V/division; middle and lower trace 5V/division; horizontal 2ms/division

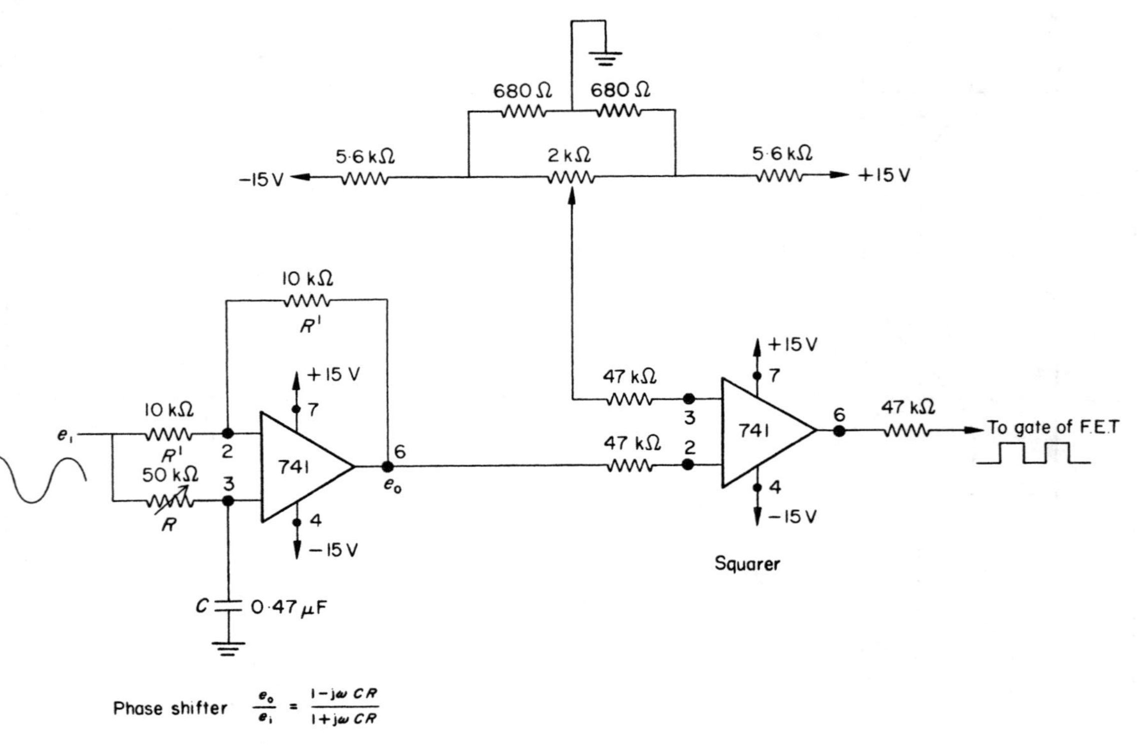

$$\text{Phase shifter} \quad \frac{e_o}{e_i} = \frac{1 - j\omega CR}{1 + j\omega CR}$$

$$\text{Phase shift} \quad \theta = -2\tan^{-1}\omega CR$$

Fig. 4.11　Phase shifting and squaring circuit

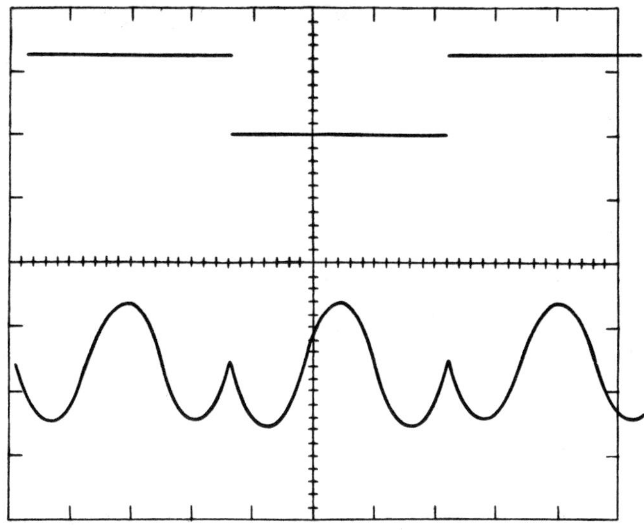

Fig. 4.12   Response of PSD to third harmonic of reference

4.3 Some Measurements on Transistors

The test circuit shown in figure 4.13 may be used to obtain a display showing the way in which the common emitter current gain, β, of a bipolar transistor, varies with V_{ce} and I_c. The operational amplifier in the circuit is connected as a differentiator and the collector emitter voltage of the transistor under test supplies the input voltage to this differentiator. The transistor load line equation gives

$$e_i = V_{ce} = V_s - I_c R_L$$

the output voltage e_0 of the differentiator is determined by the relationship

$$e_0 = - CR \frac{de_i}{dt}$$

or

$$e_0 = + CR R_L \frac{dI_c}{dt}$$

Now

$$I_c = \beta I_b$$

thus

$$e_0 = CR R_L \frac{dI_b}{dt} \beta$$

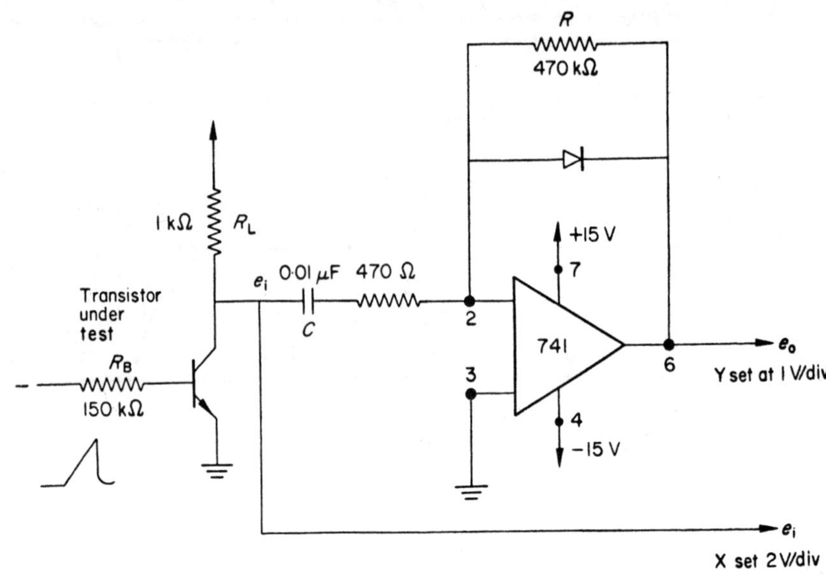

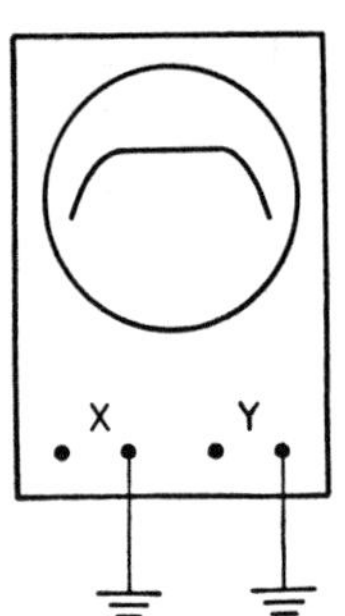

Fig. 4.13   Some measurements on transistors

In the test circuit the base current of the transistor is made to vary linearly with time by applying a linear voltage ramp to the resistor R_b and this makes dI_b/dt constant. The output voltage of the differentiator is thus directly proportional to β. It is used to provide the vertical deflection for an oscilloscope display. Horizontal deflection for the display is provided by the collecter emitter voltage of the transistor under test. Typical displays for two different transistors are shown in figure 4.14, the linear ramp used in the experiment was obtained from a second oscilloscope. The ramp had an amplitude of approximately 50 V and a rate of rise of 600 V/s making

$$\frac{dI_b}{dt} = \frac{600}{R_b} = 4 \times 10^{-3} \text{ A/s}$$

Substitution for the magnitudes of the other components used in the circuit gives

$$e_0 \cong 0.02 \ \beta \text{ volts}$$

or

$$\beta \cong 50 \ e_0$$

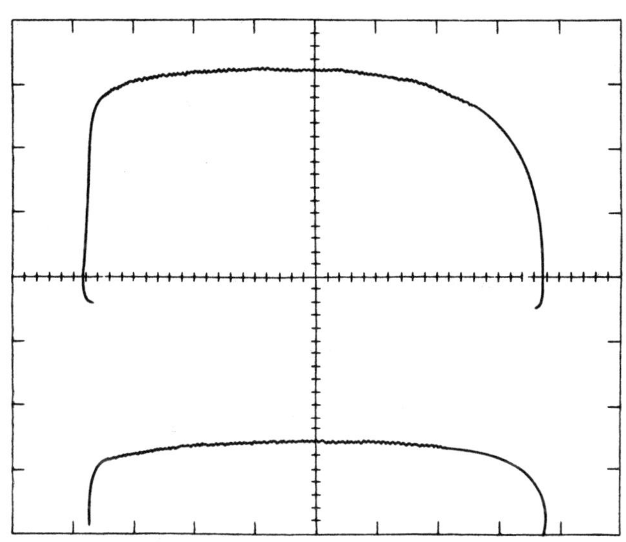

Fig. 4.14   X/Y displays showing $\beta/V_{ce} \ I_c$ for two transistors

In the oscilloscope display the vertical deflection sensitivity was set at 1 V/div and the horizontal sensitivity at 2 V/div. The trace thus corresponds to a β/V_{ce}, I_c plot with the scaling shown in figure 4.15.

   A second test circuit for transistor current gain measurement is shown in figure 4.16. This circuit is particularly suitable for measuring current gains at low values of transistor operating current. The emitter current of the transistor under test is set at a desired value by means of amplifier A_2 and its associated circuitry. Amplifier A_2 is connected as a constant current generator supplying an emitter current $I_e = - e_i/R$ where e_i is the input voltage which is applied to the constant current circuit. Amplifier A_1 is connected as a current-to-voltage converter and used to measure the base current of the transistor under test. This

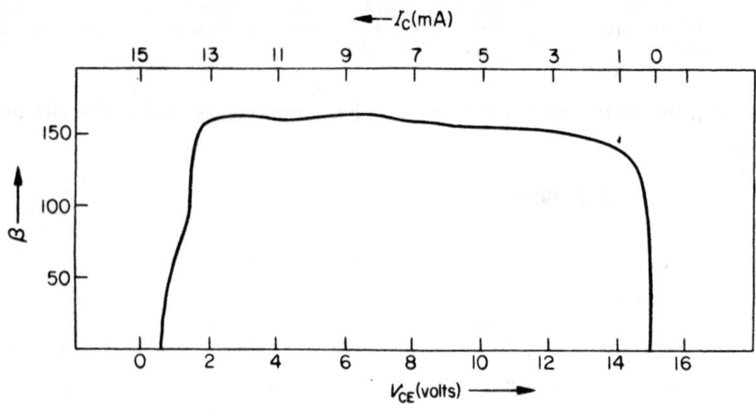

Fig. 4.15

base current is determined by the relationship $I_B = e_{01}/R_f$, where e_{01} is the output voltage of amplifier A_1. The base/emitter voltage of the transistor under test may be determined by a measurement of the output voltage of amplifier A_2 and use of the relationship, $e_{02} = -2\ V_{be}$. Before taking readings with the circuit the bias current of amplifier A_1 should be balanced. In order to make this adjustment e_i is set to zero and the potentiometer P is varied so as to make the output voltage e_{01} zero.

Two sets of readings obtained with the test circuit using an n-p-n transistor type 2N 3708 are given to illustrate the capability of the circuit.

(1) $e_i = -0.213\text{V}, \quad e_{01} = 0.1\text{V}, \quad e_{02} = -0.91\text{V}$

The readings give

$$I_e = \frac{0.213}{10^5} = 2.13 \times 10^{-6}\ \text{A}, \quad I_b = \frac{0.1}{10^6} = 10^{-7}\ \text{A}$$

$$V_{be} = -\frac{e_{02}}{2} = 0.45\text{V}, \quad \beta = \frac{I_e}{I_b} = 21.3$$

(2) $e_i = -23.6\text{V}, \quad e_{01} = 5\text{V}, \quad e_{02} = -1.23\text{V}$

The readings give

$$I_e = 236 \times 10^{-6}\ \text{A}, \quad I_b = 5 \times 10^{-6}\ \text{A}$$

$$V_{be} = 0.61\text{V}, \quad \beta = 47.2$$

Accuracy limitations are determined by resistor tolerances and by amplifier drift. The polarities of the collecter supply voltage and the input voltage e_i should be reversed if a test on a p-n-p transistor is to be made with the circuit.

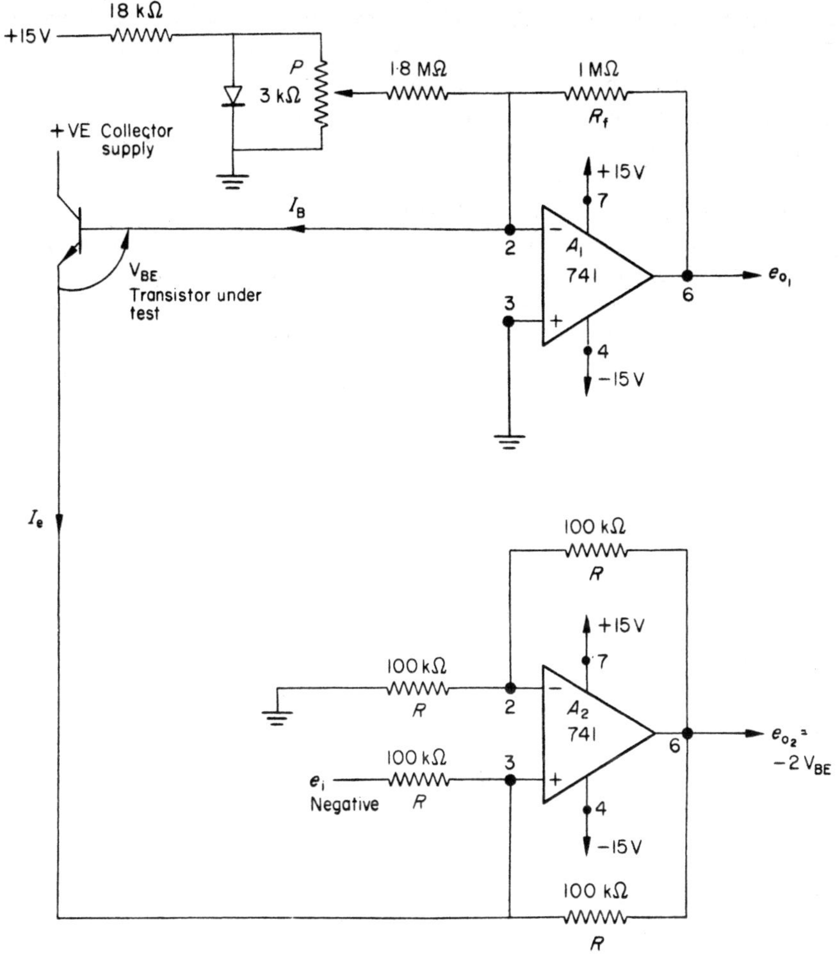

Fig. 4.16   Transistor measurements

4.4 Capacitance Measurements

The closed loop gain of an operational feedback circuit depends upon the components used in the input and feedback network. In a circuit in which the input signal is applied through a capacitor C_i and feedback is applied via a capacitor C_f the closed loop gain is determined by the ratio of the two capacitors, $A_{CL} = - C_i/C_f$. If C_f is an accurately calibrated standard capacitor, a measurement of the closed loop gain can be used as a means of accurately determining the capacitor C_i. A test circuit based upon this principle is shown in figure 4.17. The circuit incorporates the network $R_1\,C_1\,R_2$ in order to prevent the output drift which would otherwise arise because of a continuous charging of capacitor C_f caused by the amplifier bias current.

The circuit of figure 4.17 can be used to measure a wide range of capacitance values provided that in each measurement C_f is chosen so that it is of the same order of magnitude as C_i. The circuit has been used by the author to measure the values of capacitors in the range 10 μF – 10 pF using values of C_f in the range 1 μF – 100 pF. If an oscilloscope is used to monitor and measure input and output signals the oscilloscope calibrator signal can be used as a convenient test signal. Values of input capacitor greater than 0.1 μF require a test signal source of internal impedance lower than that of the oscilloscope calibrator. Measurements are insensitive to the frequency of the test signal over a considerable frequency range. Stray capacitance to earth associated with the unknown capacitor does not introduce any appreciable measurement error since the unknown capacitor is connected to the inverting input terminal of the operational amplifier and this terminal acts as a virtual earth point.

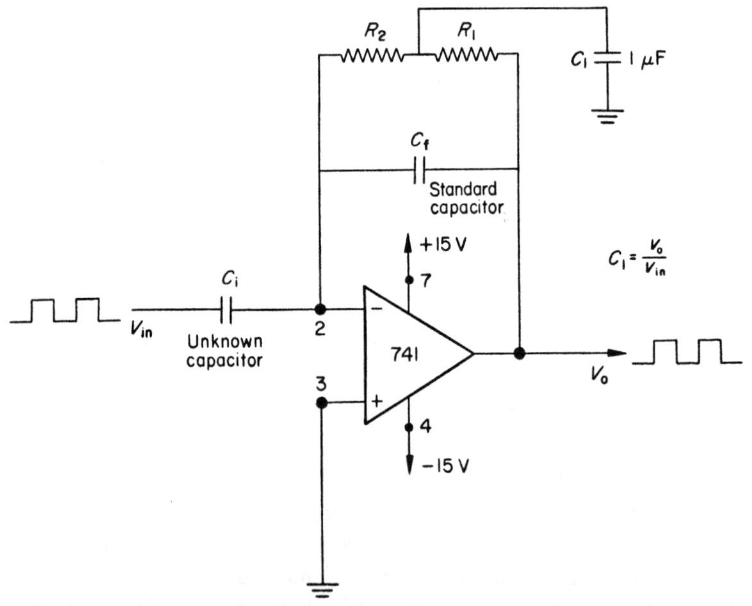

Fig. 4.17   Capacitance measurement

4.4.1 Voltage Dependence of Capacitance

The test circuit shown in figure 4.18 can be used to display and measure any voltage dependence exhibited by a capacitor. The operational amplifier in the circuit is connected as a differentiator with the capacitor under test acting as the input circuit element. A linear ramp (obtained from an oscilloscope) is used as the input test signal and the amplifier output voltage is thus proportional to the magnitude of the input capacitance

$$e_0 = - CR\ \frac{de_i}{dt}$$

The output voltage is used to provide vertical deflection and the linear ramp provides horizontal deflection for an oscilloscope display; the display represents capacitance vertically against voltage horizontally. The vertical scaling factor should be fixed at a value appropriate to the magnitude of the capacitor under test, its value is determined by choice of resistor R and ramp slope

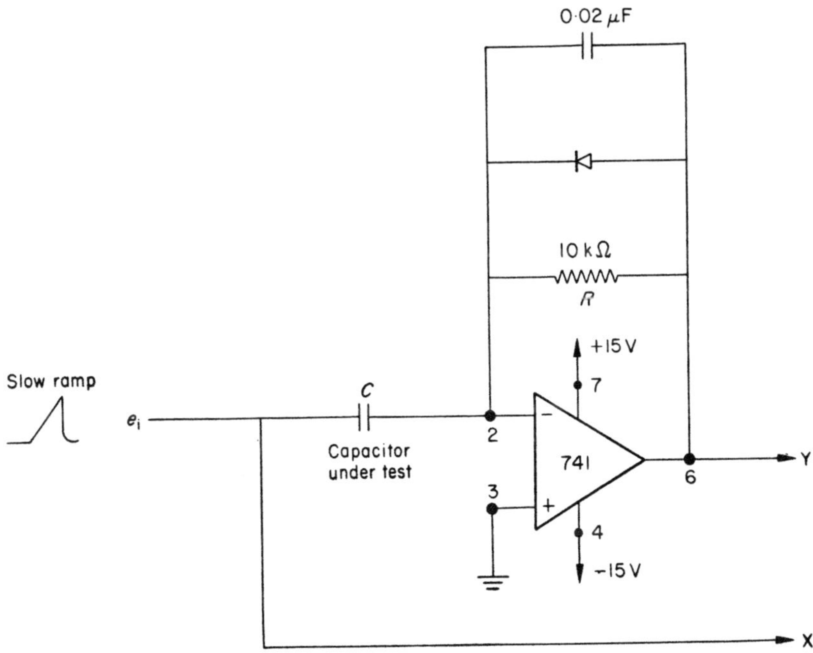

Fig. 4.18   Voltage dependance of capacitance

$$\frac{C}{e_0} = -\frac{1}{R\,\dfrac{de_i}{dt}} \quad \text{F/V}$$

Two typical oscilloscope displays are shown in figure 4.19. The upper trace
was obtained for a 10 μF and the lower trace for a 1 μF capacitor (nominal
values). Both displays are shown inverted. A linear ramp of slope 60 V/s and a
vertical deflection sensitivity of 5 V/div were used for the upper trace, making
the vertical scale for this trace 8.3 μF/div. The corresponding values for the
lower trace were 300 V/s and 2 V/div, giving a vertical scale of 0.66 μF/div. The
horizontal deflection sensitivity was 5 V/div for both traces.

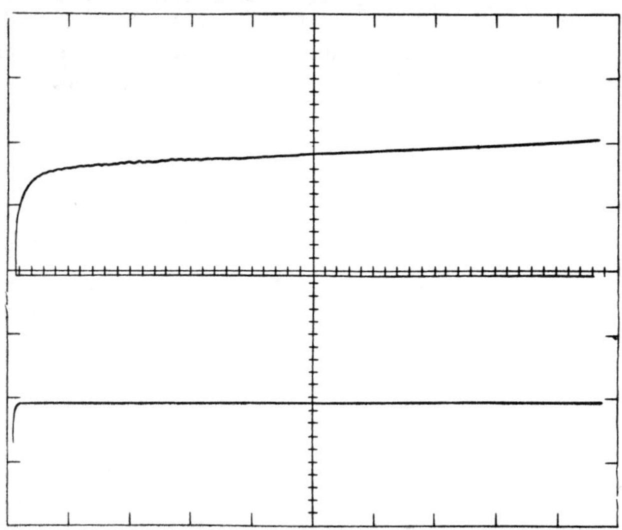

Fig. 4.19   Voltage dependance of capacitance shown by oscilloscope plot

References

1. P.C.G. Danby, Signal recovery using a phase sensitive discriminator. *Elect
   Engng*, (January 1970).

2. J.G. Lacy, *Elect Engng,* **39** (1967) 148-51.

3. P.C.G. Danby, *Elect Engng,* **40** (1968) 668-9.

4. J.L. Linsley Hood, *Elect Engng* (April 1970)

5. Operational Amplifiers used in Switching and Timing Applications

5.1 Comparators

A comparator is basically a device which compares two signals and gives an indication of which of the two is the larger. There are a variety of ways in which a differential input operational amplifier can be used to perform a comparator function. A simple circuit for demonstrating comparator action is shown in figure 5.1.

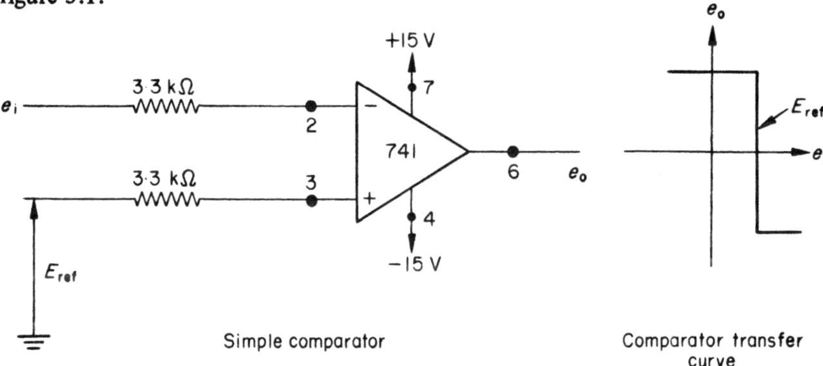

Simple comparator   Comparator transfer curve

Fig. 5.1

In this circuit the amplifier output switches between its saturation levels when the input signal applied to the phase inverting input terminal of the amplifier becomes equal to a reference signal applied to the non-inverting input terminal. Input and reference signals can be interchanged in order to obtain an output transition of reverse polarity. A second comparator circuit, in which the output transition occurs when the sum of two input voltages reaches a defined level, is shown in figure 5.2. The action of both circuits can be investigated by applying measured d.c. input signals and using a d.c. voltmeter connected to the amplifier output to indicate the state of the comparator. Alternatively a low frequency sinusoid can be used as an input signal and the comparator transfer curve can be displayed as an X/Y plot by an oscilloscope.

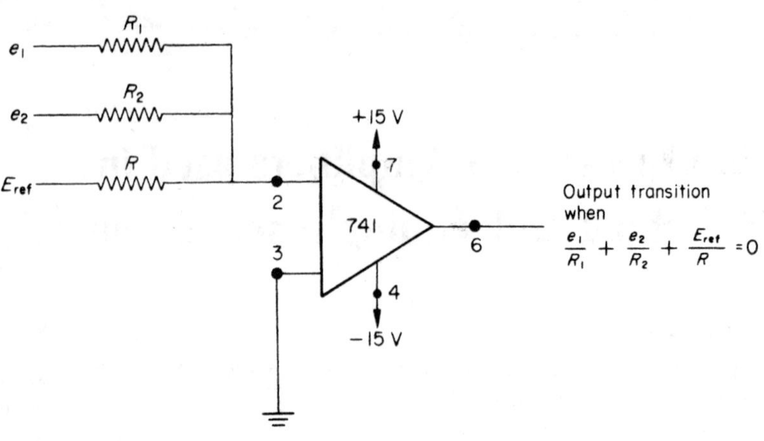

Fig. 5.2   Comparator

5.1.1 *Regenerative Comparators*

When the input signal to a simple comparator varies very slowly the comparator switching time becomes dependent upon the rate of change of the input signal. Under such circumstances comparator switching time can be reduced to a limiting value, set by amplifier slewing rate, by applying positive feedback to the amplifier. Comparators which employ positive feedback are called regenerative comparators and have a transfer curve which exhibits hysteresis. An experimental circuit for investigating the action of a regenerative comparator is shown in figure 5.3 and typical transfer curves which illustrate the action of the circuit are shown by the two traces in figure 5.4. The upper and lower transition level values of the input signal are determined by the relationships

$$e_i \bigg|_{\substack{\text{upper} \\ +ve\ \text{limit}}} = V_0 \frac{R_1}{R_1 + R_2} \qquad e_i \bigg|_{\substack{\text{lower} \\ -ve\ \text{limit}}} = V_0 \frac{R_1}{R_1 + R_2} \qquad (5.1)$$

If point B in the circuit is connected to a reference voltage supply (E_{ref}), rather than to earth as shown in the circuit, the transition levels become

$$e_i \bigg|_{\substack{\text{upper} \\ +ve\ \text{limit}}} = V_0 \frac{R_1}{R_1 + R_2} + E_{\text{ref}} \frac{R_2}{R_1 + R_2} \qquad (5.2)$$

$$e_i \bigg|_{\substack{\text{lower} \\ -ve\ \text{limit}}} = V_0 \frac{R_1}{R_1 + R_2} + E_{\text{ref}} \frac{R_2}{R_1 + R_2}$$

The equations should be checked with experimental observations and the effect of interchanging input and reference signals should also be observed.

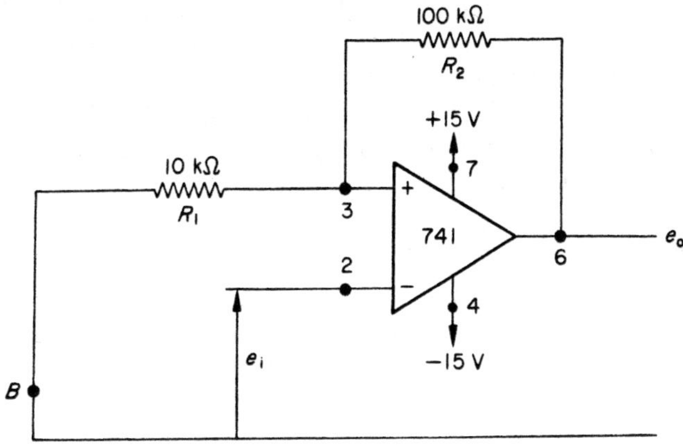

Fig. 5.3   Regenerative comparator

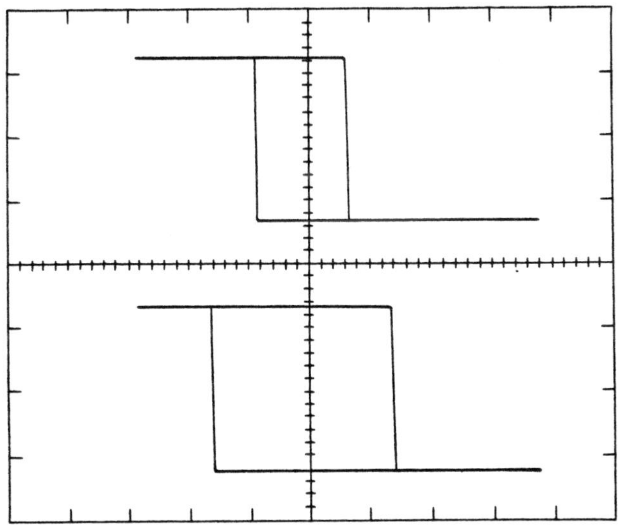

Fig. 5.4   Regenerative comparator transfer curve; vertical scale 10V/division; horizontal scale 2V/division

5.2 Multivibrator Circuits using Operational Amplifiers

Operational amplifiers are normally used in negative feedback circuits but when appropriate positive feedback connections are made to them they can be used to generate both sinusoidal and non-sinusoidal waveforms of defined frequency and amplitude. In this section we present practical circuits which illustrate ways in which positive feedback can be applied to an operational amplifier in order to give it a multivibrator type of action.

5.2.1 *Free-Running Multivibrator*

A circuit suitable for investigating the behaviour of an operational amplifier free running multivibrator is shown in figure 5.5. Positive feedback is applied to the amplifier by the resistive divider R_1 R_2. The divider gives a positive feedback fraction

$$\beta = \frac{R_1}{R_1 + R_2}$$

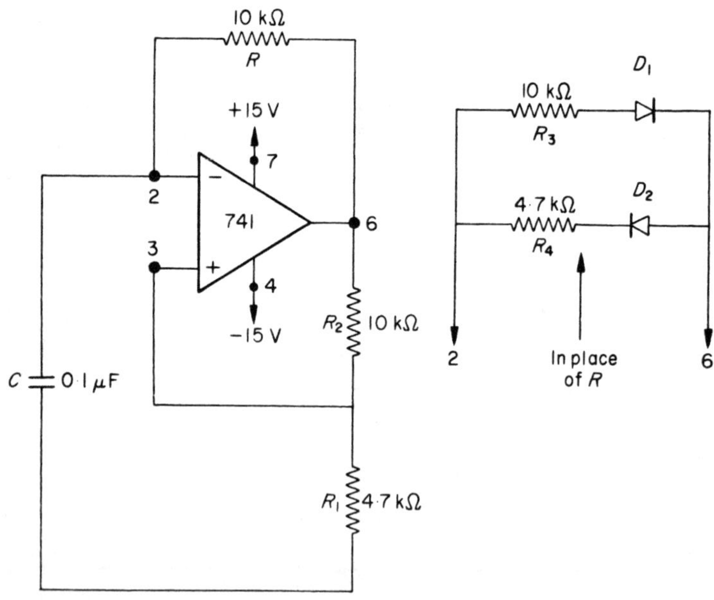

Fig. 5.5   Free running multivibrator

The amplifier switches regeneratively and repetitively between saturated states, remaining in alternate states for time periods governed by capacitor charging. The amplifier remains in positive saturation for a time period

$$t_1 \;=\; CR \;\; \log_e \frac{V^+_{0/sat} - \beta\, V^-_{0/sat}}{V^+_{0/sat} - \beta\, V^+_{0/sat}} \tag{5.3}$$

$$t_2 \;=\; CR \;\; \log_e \frac{V^-_{0/sat} - \beta\, V^+_{0/sat}}{V^-_{0/sat} - \beta\, V^-_{0/sat}} \tag{5.4}$$

The two time periods are equal and the waveforms produced are symmetrical if the positive and negative output saturation limits of the amplifier have the same magnitude.

It is suggested that the action of the circuit be investigated by observing and recording the waveforms which appear at pins 6, 3 and 2. The amplitudes and time periods of all waveforms should be measured. Typical waveforms obtained with the circuit are shown in figure 5.6. The upper trace shows the amplifier output voltage as it switches between saturation limits, the middle trace shows the signal at the non-phase inverting input terminal of the amplifier which

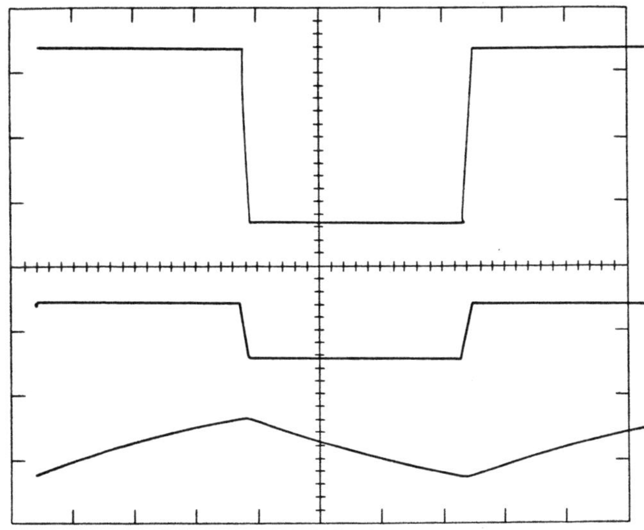

Fig. 5.6   Free running multivibrator waveforms

switches between the limits $\beta V^+_{0/sat}$ and $\beta V^-_{0/sat}$, and the lower trace shows the exponential charging of the capacitor connected to the phase inverting input terminal of the amplifier. The exponential charges up and down between the limits $\beta V^+_{0/sat}$ and $\beta V^-_{0/sat}$.

Values should be substituted into equation 5.3 and 5.4 in order to compare predicted timing periods with those obtained experimentally. A further understanding of the circuit action can be gained by changing component values and by making separate changes in the values of the positive and negative power supplies. The effect of each change on the waveforms should be noted and the reader should then attempt to explain these effects in terms of the action of the circuit.

The circuit can be made to produce a non-symmetrical waveform if the resistor R is replaced by the network, R_3 diode D_1, R_4 diode D_2; capacitor C then charges up through resistor R_4 and charges down through resistor R_3.

Control of the pulse width produced by a free running multivibrator can be obtained by externally injecting an additional current into the circuit at the phase inverting input terminal of the amplifier. The effect of this current is to increase one timing period and decrease the other. A circuit which demonstrates

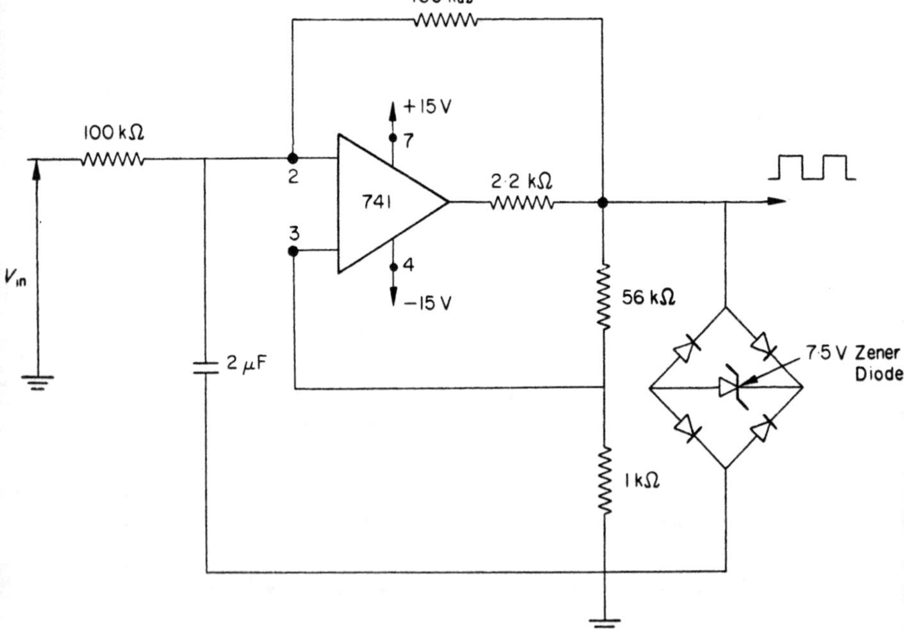

Fig. 5.7   Free running multivibrator with pulse width control and output limiting

this effect is shown in figure 5.7. The circuit in figure 5.7 includes a method for symmetrically clamping the output voltage limits using a diode bridge and zener diode. The clamp is not essential to the action of the circuit, it is included merely to illustrate a method of output limiting. Output limiting may be applied to any of the switching circuits described if the application requires it. Waveforms obtained with the circuit of figure 5.7 are shown in figure 5.8. The traces show the output voltage of the amplifier and the voltage at pin 2 for values of V_{in} equal +5V and −5V. Note that pulse width is not linearly related to the input voltage V_{in} because capacitor C charges exponentially. Linearity can be improved by reducing the amplitude of the waveform at pin 2 (by reducing the value of resistor R).

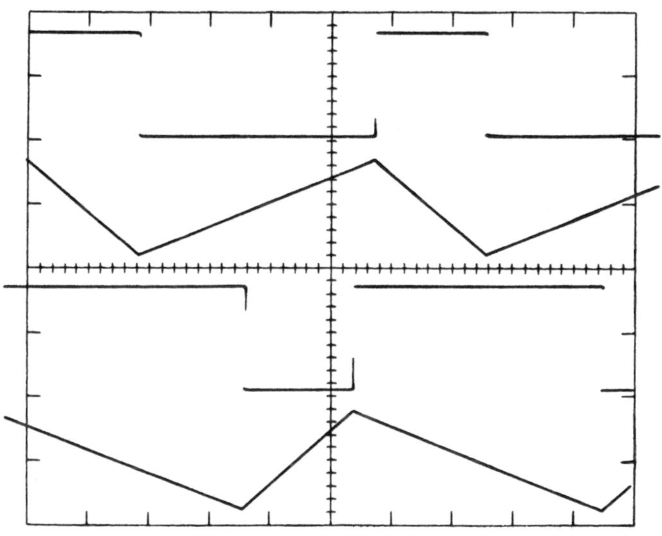

Fig. 5.8   Waveforms for free running multivibrator with pulse width control; square wave 10V/division; exponential 0.2V/division; horizontal 2 ms/division

5.2.2 *Monostable Multivibrator*

In figure 5.9 we illustrate an operational amplifier multivibrator circuit which is a monostable with a timing period controlled by the magnitude of a reference voltage. The permanently stable state of this circuit occurs when the amplifier output voltage is at its positive saturation limit. This condition is maintained by the negative reference voltage which is applied to the phase inverting input terminal of the amplifier. A positive triggering voltage applied to the phase inverting input terminal, if it is large enough to bring the amplifier out of saturation, causes the circuit to switch regeneratively to its temporarily stable

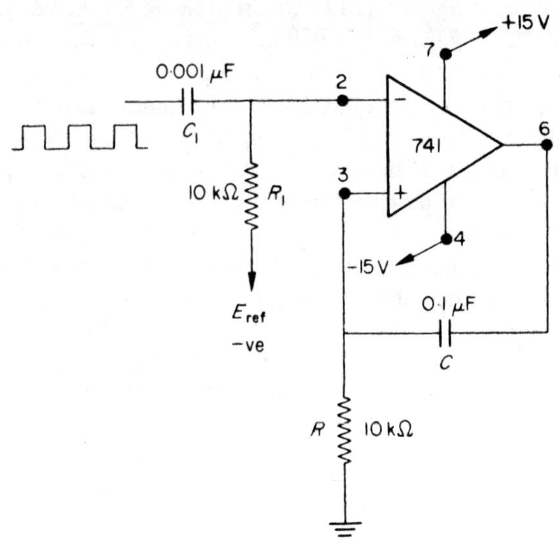

Fig. 5.9   Monostable multivibrator

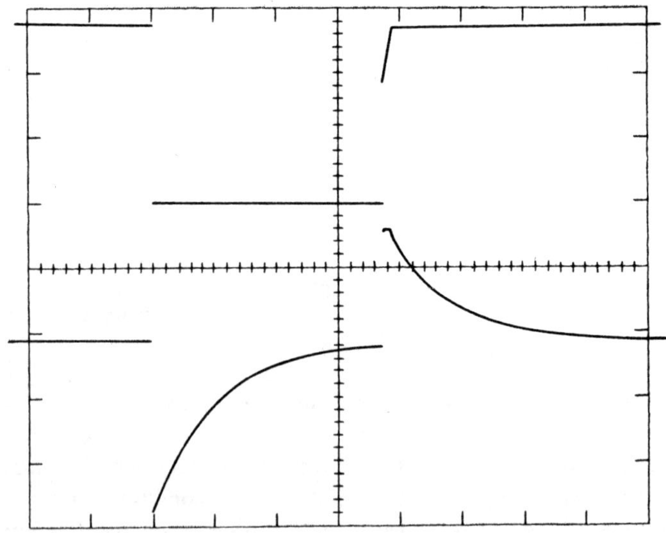

Fig. 5.10   Monostable waveforms, upper trace pin 6, lower trace pin 3; vertical scale 10V/division; horizontal scale 1 ms/division; $E_{ref} = 0.5V$

state. In this regenerative action the voltage at the non-phase inverting input terminal is switched below earth by an amount $(V^+_{0/sat} - V^-_{0/sat})$. The voltage at this terminal then rises exponentially towards earth as capacitor C charges up. The circuit switches back to its permanently stable state when this rising voltage reaches the value of the negative reference voltage. The timing period for the circuit is given by the equation

$$T = CR \log_e \frac{V^+_{0/sat} - V^-_{0/sat}}{E_{ref}} \tag{5.5}$$

The action of the circuit can be investigated by applying a square wave of amplitude, say 6V, and frequency approximately 200 Hz to the phase inverting input terminal via capacitor C_1. The square wave is differentiated by $C_1 R_1$ and the positive pulses cause the monostable to make its transition. Typical waveforms appearing at pins 6 and 3 for reference voltages −0.5 V and −5 V are shown in figures 5.10 and 5.11 respectively.

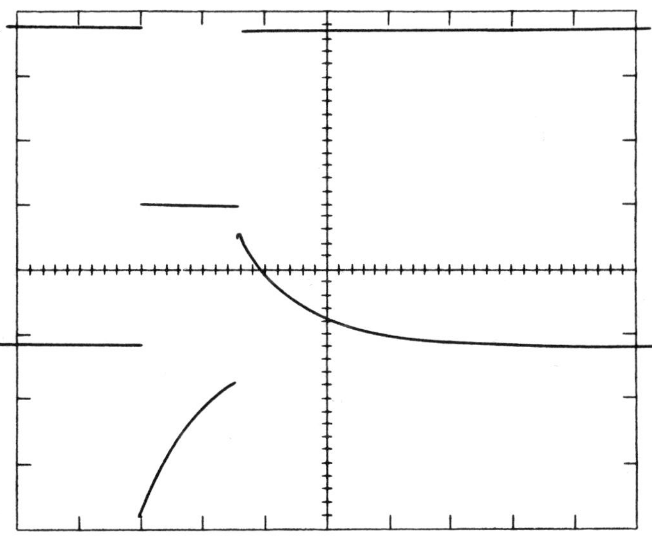

Fig. 5.11   Monostable waveforms, upper trace pin 6, lower trace pin 3; vertical scale 10V/division; horizontal scale 1 ms/division; E_{ref}= −5V

5.2.3 *Bistable Multivibrator*

A circuit which uses an operational amplifier as a bistable multivibrator is shown in figure 5.12. Positive feedback applied via resistors R_1, R_2, causes the amplifier output to remain in one or other of its saturation levels.

Triggering pulses may be applied to the circuit at either input terminal through the capacitors C_1 and C_2. Pulse polarity required to produce a transition depends upon the state of the circuit.

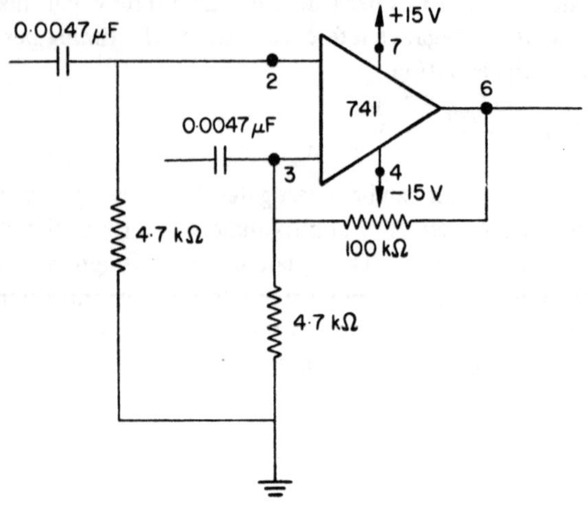

Fig. 5.12   Bistable multivibrator

5.3 Operational Amplifier Timing Circuits

In instrumentation systems it is sometimes necessary to change the nature of a signal or convert it from one form into another. Conversions which involve time, for example, voltage to time conversion or time to voltage conversion, can be performed using an operational integrator as the basis of the conversion circuit. The experimental circuits presented in this section show some of the ways in which operational amplifiers can be used for time conversions.

5.3.1 *Pulse Height to Time Conversions*

A circuit which can be used to produce a time interval which is proportional to the height of an input pulse is shown in figure 5.13. The amplifier is connected essentially as an integrator. In the absence of an input pulse the output voltage of the amplifier is held at approximately zero because of the negative feedback through diode D_3. A positive input pulse causes capacitors C_1 and C_2 to charge; the output of the amplifier steps down and diode D_3 is reverse biased. If V_p is the height of the input pulse the magnitude of the output step is $V_p\, C_1/C_2$, this step is discharged by the integrator at a rate $E_{ref}/C_2\, R$ volts/s. The time for the output to charge back up to zero is thus

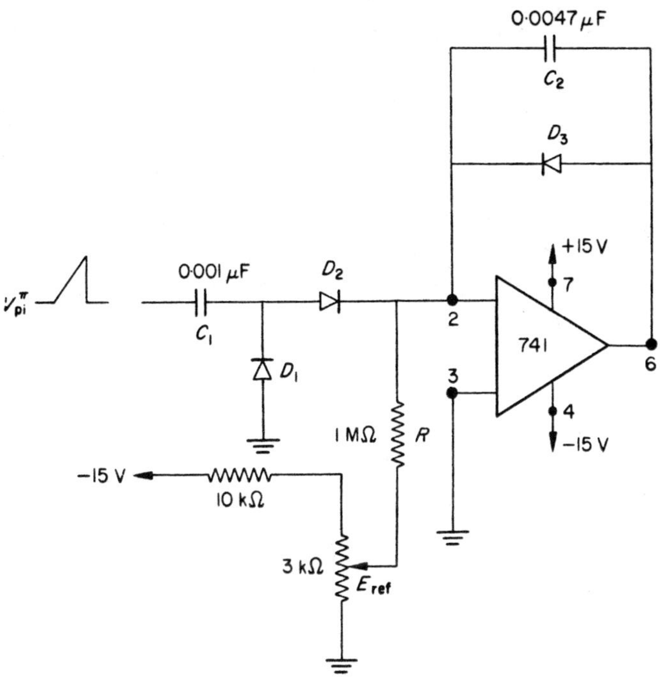

Fig. 5.13   Pulse height to time conversion

$$t = \frac{Vp\ \dfrac{C_1}{C_2}}{\dfrac{E_{ref}}{C_2\,R}} = \frac{C_1\,R}{E_{ref}} \times Vp \tag{5.6}$$

The time period is directly proportional to the height of the input pulse and scaling can be set by choice of E_{ref}.

Circuit performance can be checked by applying input pulses of different amplitude and measuring charging times by observation of oscilloscope waveforms; typical waveforms for two different input pulse heights are shown in figure 5.14.

5.3.2 Time to Voltage Conversion

A modification to the circuit or figure 5.13 may be made in order to allow a time to voltage conversion; the modification is shown in the circuit of figure

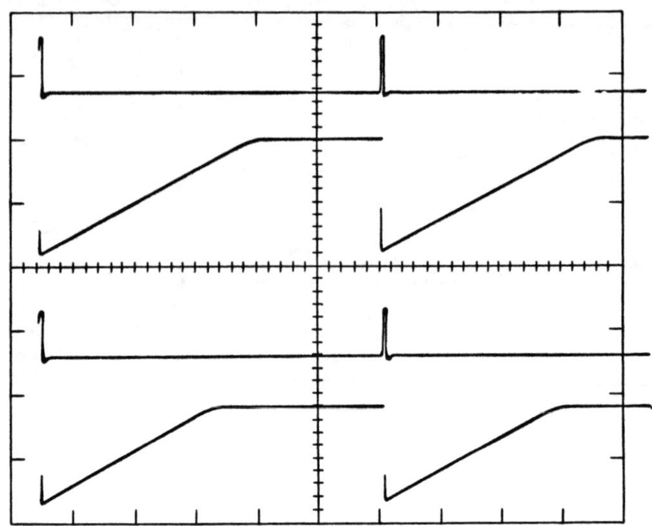

Fig. 5.14   Pulse height to time conversion; pulse waveform 10V/division; integrator output waveform 1V/division; horizontal scale 1 ms/division

5.15. The time interval to be converted is the time period of a negative gating signal. In the absence of the gating signal a positive voltage at point A causes conduction through diode D_1 and the output of the amplifier is held at approximately zero because of the negative feedback through diode D_2. A negative gate applied at point A reverse biases both diodes and the output voltage of the amplifier rises linearly at a rate determined by E_{ref} and the integrator time constant CR. The amplitude of the output ramp generated is directly proportional to the time period of the gating signal

$$E_0 = \frac{R_{ref}}{C\,R} \times t \tag{5.7}$$

E_0 is the amplitude of the output ramp and t is the time period of the gating signal.

Typical waveforms for two different gating periods are illustrated in figure 5.16.

5.3.3 D.C. Voltage to Time Conversion

A second amplifier, acting as a comparator, added to the circuit of figure 5.15 gives a system which can be used to produce a gate waveform of time period

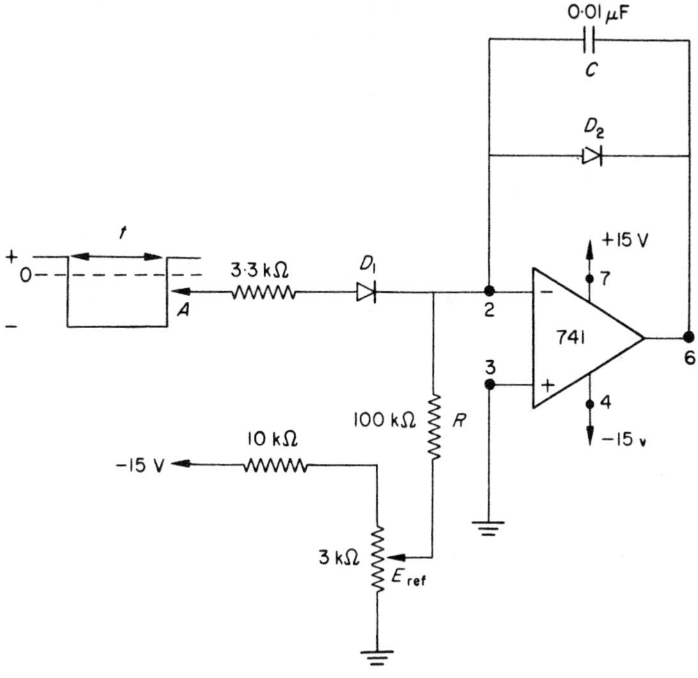

Fig. 5.15   Time to voltage conversion

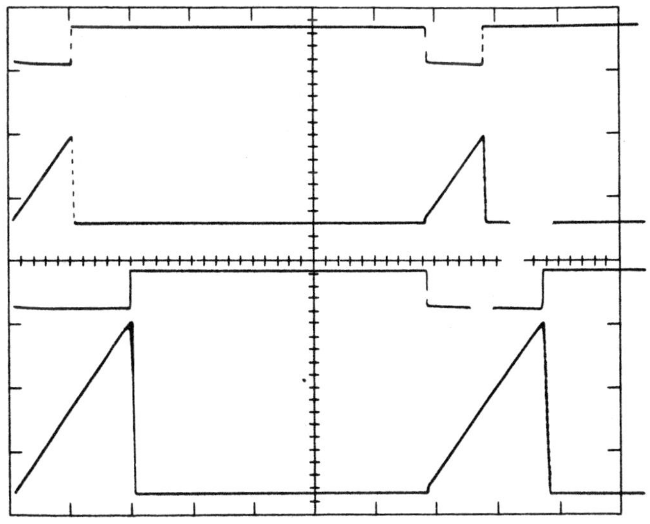

Fig. 5.16   Time to voltage conversion; square wave 5V/division; ramp 2V/division; horizontal 2 ms/division

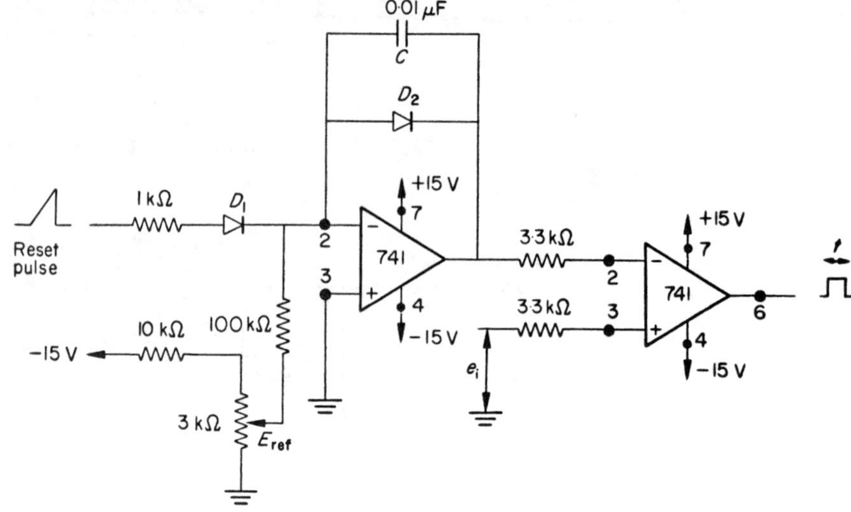

Fig. 5.17   Voltage to time conversion

directly proportional to the magnitude of a d.c. voltage. The arrangement is illustrated in figure 5.17. In this system the comparator output remains at its positive saturation level for a time period which is directly proportional to the magnitude of a positive d.c. voltage e_i , applied to the non-phase inverting input terminal of the comparator. A time conversion is initiated by a positive reset pulse which sets the integrator output to approximately zero and causes the comparator output to switch to its positive level. At the end of a reset pulse the integrator output runs up linearly at a rate determined by E_{ref} and the integrator time constant CR. The comparator senses when the integrator output reaches the level e_i and switches back to its negative level. Time periods are determined by the relationship

$$t = \frac{CR}{E_{ref}} \times e_i \qquad\qquad (5.8)$$

Equation 5.8 shows that the circuit gives a time conversion which is proportional to the ratio of two voltages, the ratio e_i/E_{ref}. Typical waveforms obtained with the system are shown in figure 5.18. It is instructive to measure timing periods for different values of the circuit conditions in order to test the validity of equation 5.8.

The input voltage e_i, in figure 5.17, must be positive and its magnitude must be less than the integrator positive output voltage swing. A circuit which can be

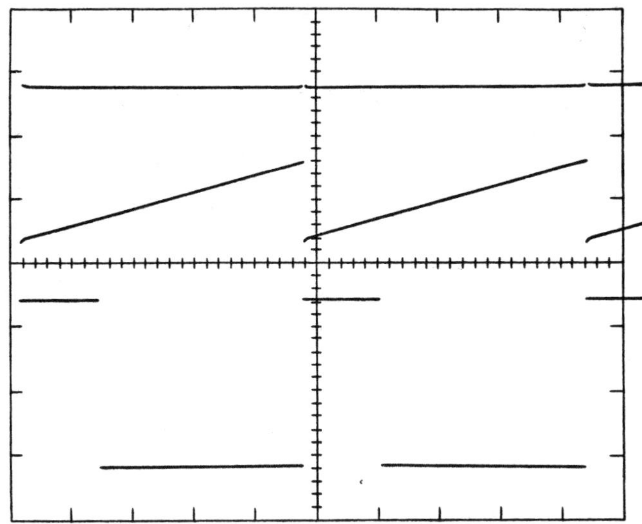

Fig. 5.18　Voltage to time conversion waveforms; upper trace, reset pulse; middle trace integrator output; lower trace, comparator output; vertical scale 10V/division; horizontal scale 2ms/division

used to provide conversion of either positive or negative input signals can be made by removing diode D_2 from the circuit so as to allow the integrator output to be reset to a negative value and applying the integrator output to two comparators. One comparator is used to sense when the integrator output passes through zero the other senses when it passes through a value equal to the input voltage. The time period between the transitions of the two comparators is then directly proportional to the magnitude of the input voltage and the sign of the input voltage is indicated by which comparator makes a transition first. A system of this kind can be used as a basis for a simple ramp type DVM, the comparator transitions being used to generate start and stop pulses which are applied to a digital timer system.

5.3.4　Voltage to Frequency Conversion

The timing function performed by an operational integrator can be used as the basis of a voltage to frequency conversion system. A voltage to frequency convertor is a system which is used to generate a pulse sequence whose repetition frequency is directly related to the magnitude of a d.c. voltage. Ideally the relationship between voltage and frequency should be linear.

An experimental circuit which uses a combination of an operational integrator and a regenerative comparator to perform a voltage to frequency

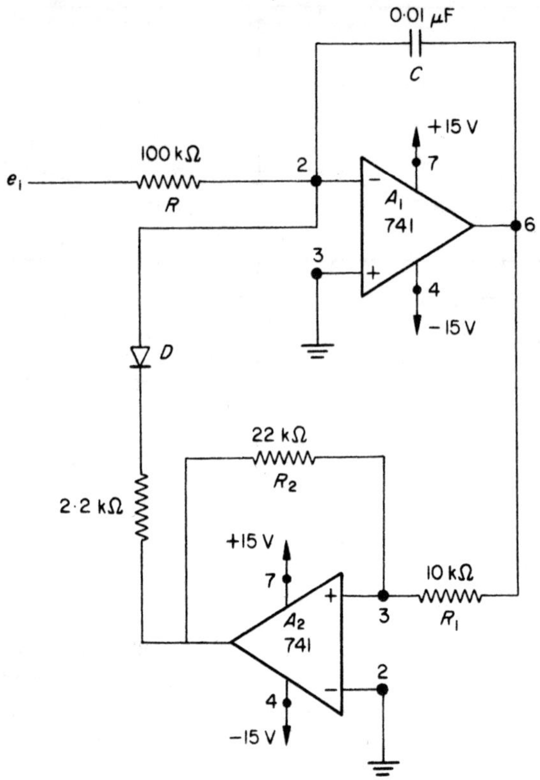

Fig. 5.19   Voltage to frequency conversion

conversion is shown in figure 5.19. Starting at the time at which the comparator output switches to its positive level, $V_{0}^{+}{}_{\text{sat}}$, the action of the circuit, is as follows. Diode D is reverse biased and the output of the integrator falls linearly at a rate determined by the magnitude of the positive d.c. input voltage e_i . When the integrator output reaches a level $- V_{0_{\text{sat}}} R_1/R_2$ the comparator switches to its negative output state, diode D is forward biased and integrator output is made to run up rapidly. When it reaches a level $- V_{0_{\text{sat}}} R_1/R_2$ the comparator reverts to its positive output state and the cycle repeats. If the time taken for the integrator output to run up is made negligibly small compared to the run down time the frequency of oscillations becomes directly proportional to e_i . An expression for the frequency of oscillations obtained by neglecting integrator run up time and comparator switching time is

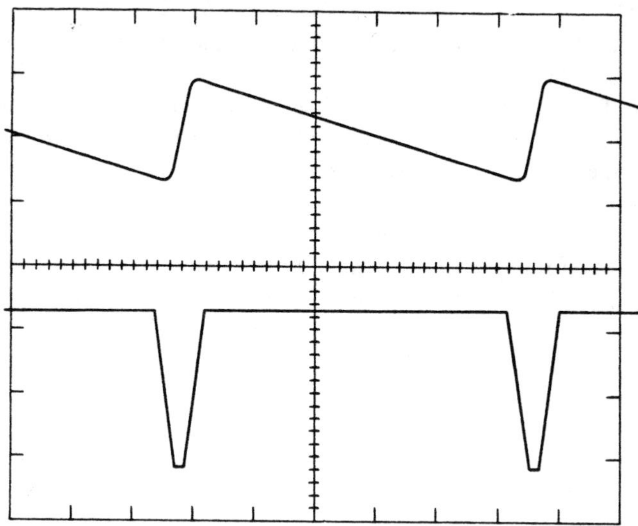

Fig. 5.20　Voltage to frequency conversion waveforms; upper trace, integrator output; lower trace, comparator output; vertical scale 10V/division; horizontal scale 0.1ms/division

$$f = \frac{e_i}{CR} \; \frac{R_2}{R_1 \, (V_{0_{sat}}^+ - V_{0_{sat}}^-)} \tag{5.9}$$

Typical waveforms appearing at the output of each amplifier are shown in figure 5.20. An applied input voltage larger than that for which the circuit converts linearly was used in order to make visible the effect of the finite comparator switching time. The graticule line cutting across the middle of each trace represents the d.c. zero level of each trace, a close inspection of the waveforms reveals the d.c. levels at which switching occurs. Notice that the finite comparator switching time allows an integrator output swing which is larger than $R_1/R_2 \, (V_{0_{sat}}^+ - V_{0_{sat}}^-)$ and the frequency of oscillations is thus less than that predicted by equation 5.9.

It is suggested that the range of linear operation for the converter be examined by applying various input voltages and measuring the frequency of oscillation for each value of input voltage. Input voltages in the range 10 mV to 20 V are suggested. The addition of an offset balance to the integrator and the cancellation of integrator drift will be found to extend the lower frequency limit for linearity of voltage to frequency conversion.

Exercises 5

5.1   In the regenerative comparator circuit shown in figure 5.3 the
      inverting input terminal of the amplifier is connected to earth.
      If $R_1 = 5$ kΩ and $R_2 = 10$ kΩ and the output limits of the amplifier
      are $\pm$ 10 volts find the values of an input signal applied to R_1 for which
      the circuit switches state. Sketch the transfer curve.

5.2   Using the circuit of figure 5.5 it is required to produce an asymmetrical
      waveform frequency 1 kHz and mark space ratio of 3 : 1. If the output
      saturation limits of the amplifier are $\pm$ 10 volts and the sawtooth
      waveform is to have an amplitude of 10 volts peak to peak find suitable
      values for the components in the circuit. Sketch waveforms.

5.3   In the circuit of figure 5.13E_{ref} is set at 1 volt. Calculate the
      theoretical value of the voltage to time conversion factor for the circuit.
      Discuss the factors which limit conversion accuracy in the circuit.

5.4   Repeat question 5.3 for the circuit of figure 5.15.

5.5   A system is required which will produce a time interval proportional to
      the magnitude of a d.c. voltage which may be either positive or
      negative. Sketch an outline of a system suitable for this purpose with
      waveforms illustrating its action; consider the factors which will limit
      accuracy (figure 5.17 modified).

5.6   The output limits of the amplifiers used in figure 5.19 are $\pm$ 10 volts.
      Calculate the theoretical value of the conversion factor for the system
      if the finite switching time and finite integrator run up times are
      neglected. What is the effect of finite amplifier switching times, does it
      effect the linearity of voltage to frequency conversion?

      If the amplifier used in the integrator circuit has the values of V_{io} and
      I_b given in Exercises 2, for what value of the input voltage can
      integrator drift result in 5 per cent error?

6. Operational Amplifiers used for Signal Generation

This chapter introduces practical circuits showing some of the ways in which operational amplifiers can be used for signal generation.

6.1 Sinusoidal Oscillators

An operational amplifier can be used to produce a sinusoidal signal if external components which provide an appropriate positive feedback path are connected to the amplifier. The circuit thus formed constitutes a feedback oscillator. The condition for oscillations to start and grow in a feedback oscillator circuit is that the gain round the feedback loop should be real and positive, (the physical meaning of this mathematical condition is a zero resultant phase shift), and greater than unity. In order that waveforms produced should be sinusoidal and of defined frequency and amplitude, feedback oscillator circuits are designed to make the loop gain dependent upon both the frequency and amplitude of any signal present in the loop. Loop gain is made greater than unity for small signal amplitudes, ensuring that oscillations will start and grow in amplitude. The circuit is so designed that the loop gain becomes exactly unity for oscillations at the desired frequency and amplitude, thus ensuring the maintenance of the desired oscillation.

6.1.1 *Wien Bridge Oscillator*

The Wien bridge oscillator is a well-known form of feedback oscillator. The name arises from the type of network employed in the feedback path. The amplifier used to activate this network may be a valve amplifier, a discrete component transistor amplifier or a monolithic circuit; operational amplifiers are well suited for use in Wien bridge oscillators.

An operational amplifier form of Wien bridge oscillator is illustrated in figure 6.1. The circuit is particularly convenient as a means of investigating the feedback oscillator principles previously mentioned. There are two feedback paths in the circuit, positive feedback via impedances Z_2 and Z_1, and negative feedback via resistors R_2 and R_1. The resultant feedback fraction is

$$\beta = \frac{Z_1}{Z_1 + Z_2} - \frac{R_1}{R_1 + R_2}$$

If A_{0L} is very large and positive, substitution for the impedance values shows that the condition for oscillations, loop gain equal to $\beta A_{0L} = 1$, is satisfied at a frequency $f = 1/2\pi CR$ for values of resistors R_1 and R_2 such that R_1 is slightly less than $R_2/2$.

In the circuit of figure 6.1, R_1 is made up of two resistors, R_1' and R_{ds}. R_{ds} is the drain source resistance of the field effect transistor T_1. The FET is used as a means of ensuring stability of oscillation amplitude. If the drain source voltage is small an FET behaves like a linear variable resistor, the magnitude of R_{ds} varying almost linearly with the gate source voltage V_{gs}. R_1' is chosen so that when $V_{gs} = 0$, R_1 is appreciably less than $R_2/2$. This makes the loop gain greater

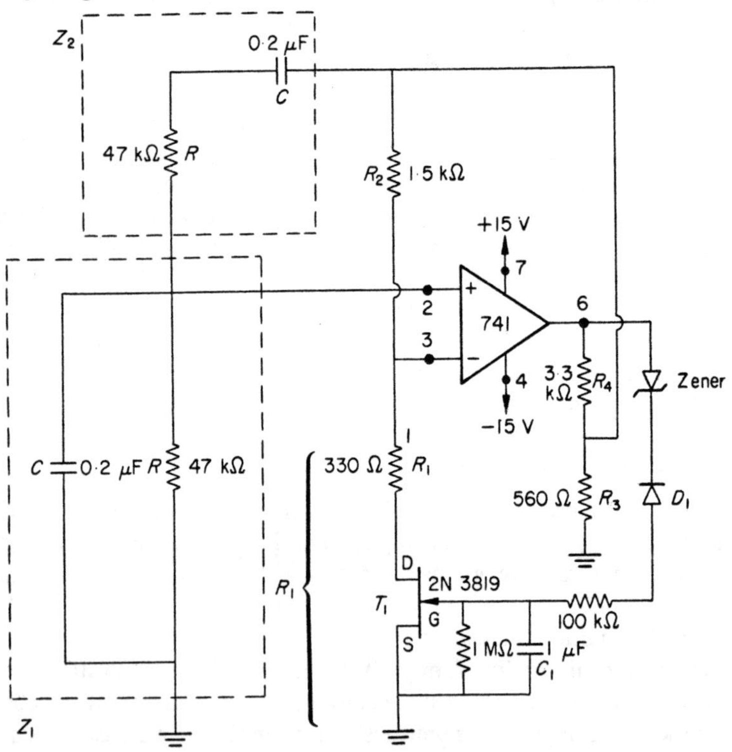

Fig. 6.1   Wien Bridge oscillator

than unity and oscillations start and grow. When the amplitude of oscillation exceeds the zener diode breakdown voltage the negative peaks of the output signal are rectified by the diode D_1. The rectified and smoothed d.c. is applied as a negative bias to the gate of the FET. The negative bias causes an increase in R_{ds} which, in turn, causes an increase in the negative feedback and a consequent reduction in the loop gain. V_{gs} eventually takes on that value which is required to make the loop gain exactly unity and so a stable amplitude of oscillations is maintained. The output signal of the amplifier is attenuated by resistors R_3 and R_4 before being applied to the feedback network. This is to ensure that the voltage between drain and source of the FET does not become great enough to cause operation outside its linear resistance region. Non-linear operation of the FET causes waveform distortion.

It is instructive to observe the effects of changing circuit conditions on the circuit waveforms at the start of oscillations. The waveforms at the amplifier output terminal and at the gate of the FET should be observed, oscillations may be stopped and started by breaking and then reconnecting the positive feedback loop. FET parameters are subject to a considerable spread, the value of the resistor used for R_1' will require selection to make it appropriate to the R_{ds} value of the particular FET used in the circuit. The effect of the value of R_1' on the rate at which the oscillation amplitude builds up should be examined, as should the effect of changing the time constant $C_1 R_5$. This time constant must be appropriate for the oscillation frequency, if it is too long oscillations will be intermittent, if it is too short oscillation waveforms will be distorted.

6.1.2 *Quadrature Oscillator*

The circuit illustrated in figure 6.2 uses two operational amplifiers connected in a feedback loop so as to produce two sinusoidal signals, one exactly 90° out of phase with the other. Amplifier A_1 acts as a non-inverting integrator and amplifier A_2 acts as an inverting integrator. The circuit may be regarded as an analogue computing loop designed to solve a differential equation of the form

$$\frac{d^2 x}{dt^2} = - \omega^2 x$$

A solution to this equation is $x = a \sin \omega t$ with the angular frequency ω determined by the time constants of the two integrators and given by the equation

$$\omega = \frac{1}{(C_1 R_1 C_2 R_2)^{\frac{1}{2}}}$$

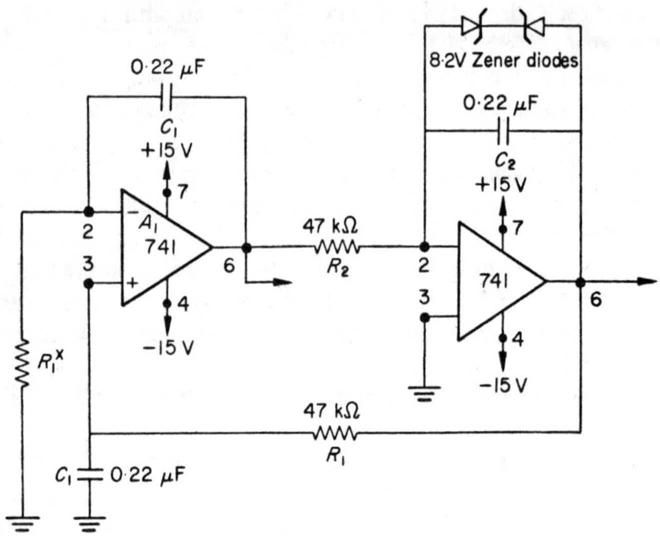

Fig. 6.2   Quadrature oscillator

In order that oscillations should start and grow in the circuit the value of
resistor R_1^x is made slightly less than R_1. Zener diodes connected in parallel with
C_2 are used to limit the amplitude of oscillations and they do this at the
expense of some distortion in waveform. The smaller value of R_1^x the more
rapidly oscillations start but the greater is the waveform distortion. A rapid build
up of oscillations without appreciable waveform distortion can be obtained by
using an FET to stabilise the amplitude of oscillations instead of the zener diodes.
A circuit in which this technique is employed is shown in figure 6.3.

6.2 Function Generator Systems

A function generator is a signal source which produces various wave shapes. The
variety of wave shapes available is normally dependent upon the complexity of
the generator system. Facilities such as voltage control of frequency and the
ability to generate a single wave or group of waves (triggered or gated operation),
are sometimes made available. In addition to their use in electronics laboratories,
function generators are applied in many other fields, such as, geophysics,
biophysics and in education where they function as simulators, testers and timers.
Function generator systems can be synthesised using operational amplifiers and
experimental arrangements for investigating such syntheses are described in this
section. In many cases the systems described consist essentially of combinations
of individual circuits which have been investigated in previous chapters.

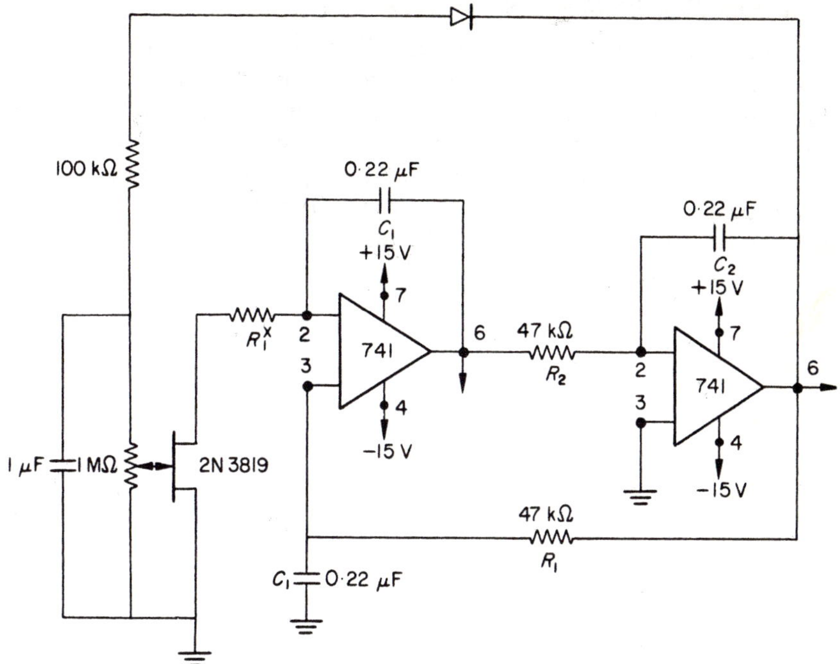

Fig. 6.3   Quadrature oscillator with FET amplitude stabilisation

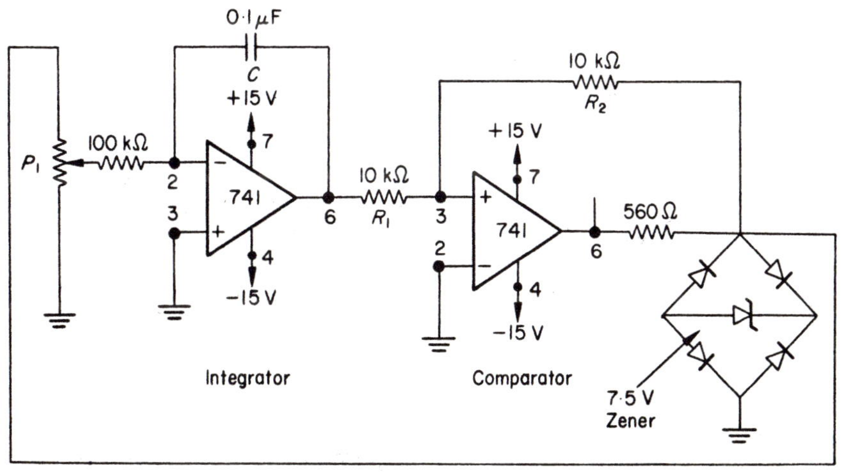

Fig. 6.4   Basic function generator

6.2.1 Basic Function Generator

The circuit for a basic function generator is illustrated in figure 6.4 and the waveforms produced by the system are shown in figure 6.5. The circuit consists of an operational integrator and a regenerative comparator connected together in a closed feedback loop. Precise triangular waves are formed by integration of the square waves fed back from the output of the comparator to the integrator input. The triangular wave amplitude is set by the input transition levels of the comparator which are determined by the relationships

$$V_t^+ = -V_0^- \frac{R_1}{R_2}$$
$$V_t^- = -V_0^+ \frac{R_1}{R_2} \tag{6.1}$$

V_t^{+-} are the upper and lower input transition levels of the comparator.

V_0^{+-} are the upper and lower output levels of the comparator. In figure 6.9 the output levels of the comparator are limited by the zener diode and diode bridge which act as a symmetrical clamp.

The frequency of the oscillations produced by the basic function generator circuit is determined by the integrator time constant and by the amplitude of the square wave fed back to the integrator input (controlled by potentiometer P_1). The time taken for the integrator output to run up is

$$T_1 = \frac{V_t^+ - V_t^-}{k \, V_0^-} \, CR \tag{6.2}$$

and the time for the output to run down is

$$T_2 = \frac{V_t^+ - V_t^-}{k \, V_0^+} \tag{6.3}$$

k has a value which can be set between zero and unity by potentiometer P. If the magnitude of the upper and lower output levels of the comparator are made the same, $T_1 = T_2$, the waveform is symmetrical and the frequency of oscillations is theoretically determined by the relationship

$$f = k \, \frac{R_2}{4 \, f_1 \, CR} \tag{6.4}$$

In a practical function generator it is usual to range switch the frequency by switching in different integrator CR values and to use the potentiometer P for control of frequency within a particular range.

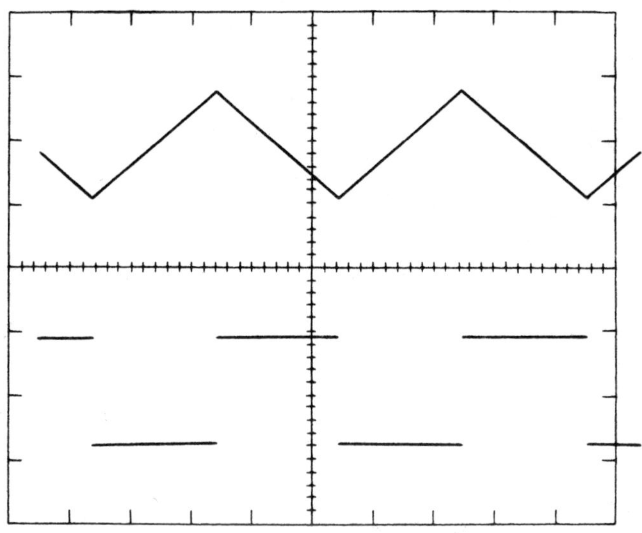

Fig. 6.5   Waveforms produced by basic function generator; vertical 10V/division;
horizontal 2ms/division

It is suggested that the reader investigate the action of the circuit using a
variety of component values, it is instructive to find the range of frequencies for
which equation 6.4 is valid. It will be found that the upper limit is determined by
amplifier slewing rate and the lower limit by integrator drift. Integrator drift rate
is a function of amplifier offsets and integrator capacitor/resistor values. If initial
offsets are balanced the offset drift performance of the particular amplifier type
used as an integrator sets an ultimate limit to the lowest frequency which can be
generated.

The circuit of figure 6.4 can be made the basis of more complex function
generator systems. A simple modification which changes the time symmetry of
the waveforms is shown in figure 6.6. A marked change in time symmetry results
in a ramp waveform at the integrator output and a pulse at the output of the
comparator. In figure 6.6 a precision rectifier has been added to the system to
rectify the ramp waveform to give either a positive or negative-going ramp.
Typical circuit waveforms with the diode D_1 switched into the circuit are shown
in figure 6.7. Waveforms with D_2 in circuit are shown in figure 6.8.

     88 PRACTICAL OP AMP CIRCUITS YOU CAN BUILD

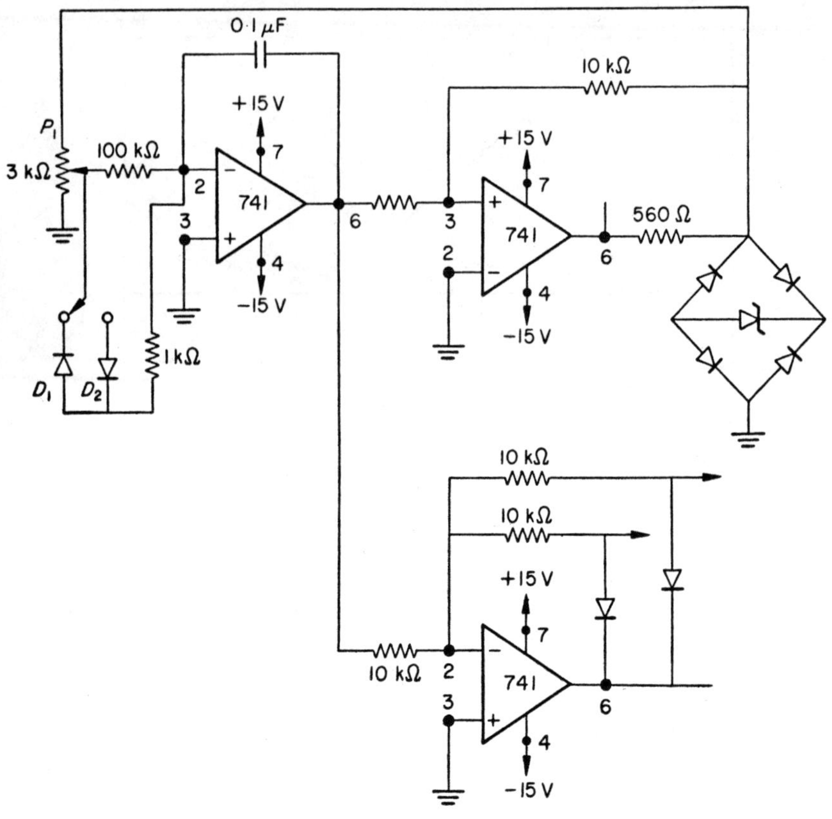

Fig. 6.6   Function generator with non-symmetrical waveforms

6.2.2 *Sine Shaping*

It is often convenient to have a sinusoidal waveform as one of the wave shapes
produced by a function generator. A sinusoidal wave can be generated from a
triangular wave by an appropriate shaping operation. An operational amplifier
with a suitably designed non-linear response can be used to perform such a
shaping and a circuit which can be used to investigate the technique is shown in
figure 6.9. In this circuit, as the amplifier output signal increases in amplitude
the diode resistor networks in the feedback path are used progressively to reduce
the closed loop gain of the amplifier by returning current to the amplifier
summing point. The process is illustrated by the waveforms shown in figure 6.10.

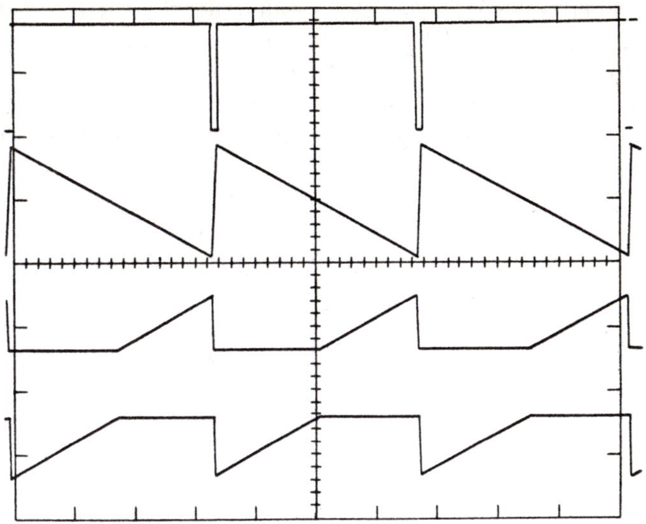

Fig. 6.7   Waveforms of figure 6.6; vertical 10V/division; horizontal 2ms/division

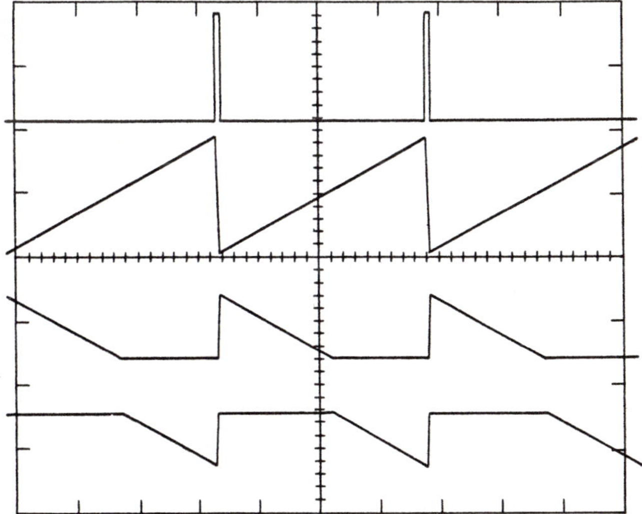

Fig. 6.8   Waveforms of figure 6.6; vertical 10V/division; horizontal 2ms/division

The top trace shows the output waveform with feedback through resistor R alone, the other traces show what happens to this waveform as diodes; D_1 D_1' D_2 D_2' D_3 D_3' and their associated resistor networks are successively connected into the circuit. A sinusoid is, in effect, approximated by a series of straight line increments, the greater the number of increments the closer is the approximation to a true sinusoid.

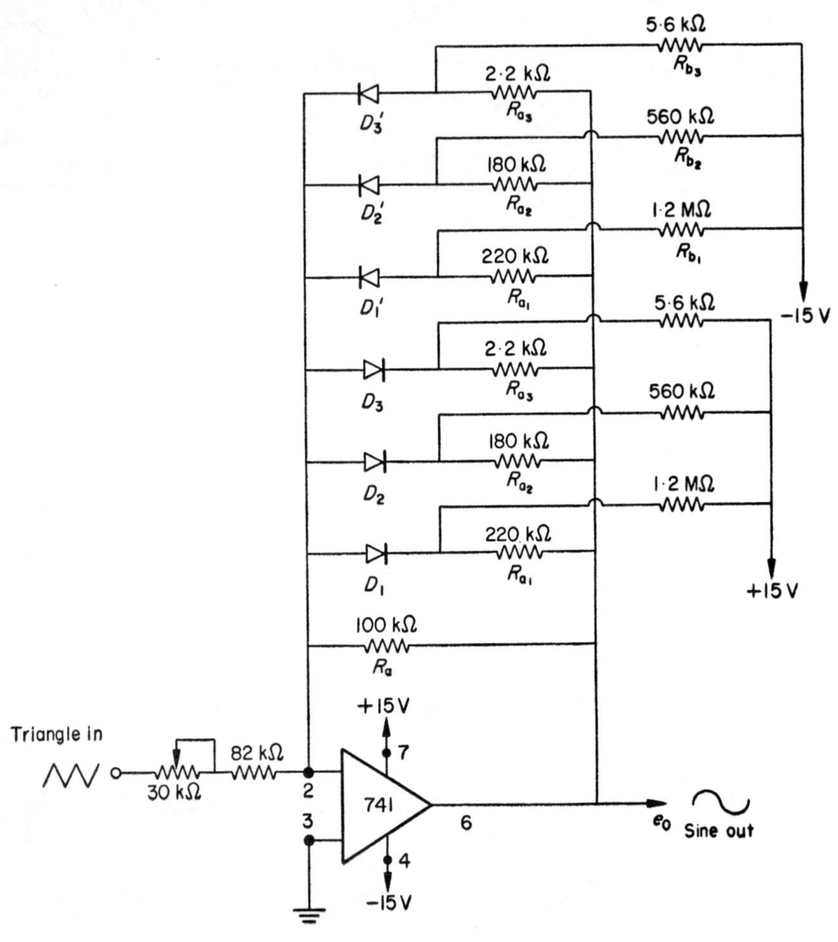

Fig. 6.9   Sine shaping circuit

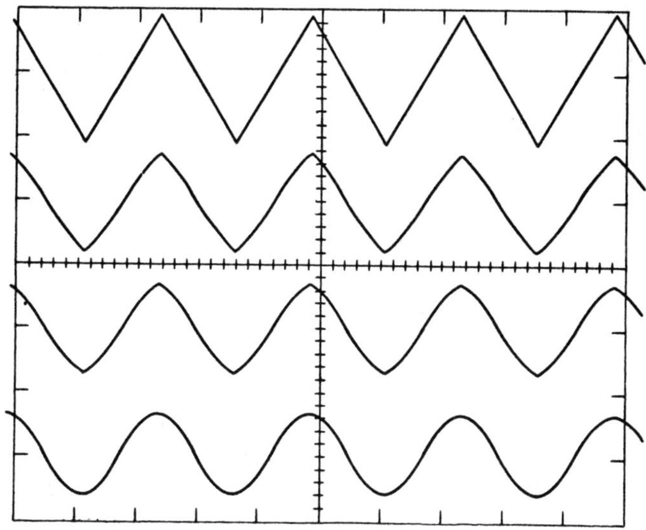

Fig. 6.10 Waveforms illustrating action of sine shaping circuit

The required slopes for the straight line increments and the break voltages can be found graphically as shown in figure 6.11. A triangular wave of period T and amplitude A has a slope $4A/T$. An expression for a sinusoidal wave with the same period is, $v_s = V_s \sin 2\pi t/T$, with a maximum slope $2\pi V_s/T$. In order that both waves should have the same slope as they pass through zero

$$\frac{4A}{T} = \frac{2\pi V_s}{T}$$

and

$$V_s = \frac{2A}{\pi} = 0.635\,A$$

Values of the resistors R_a required experimentally to set the desired slopes of the straight line increments are determined by the equations

$$\frac{\text{Desired slope of increment 1}}{\text{Slope of triangular wave}} = \frac{R//R_{a1}}{R} \tag{6.5}$$

$$\frac{\text{Desired slope of increment 2}}{\text{Slope of triangular wave}} = \frac{R//R_{a1}//R_{a2}}{R}$$

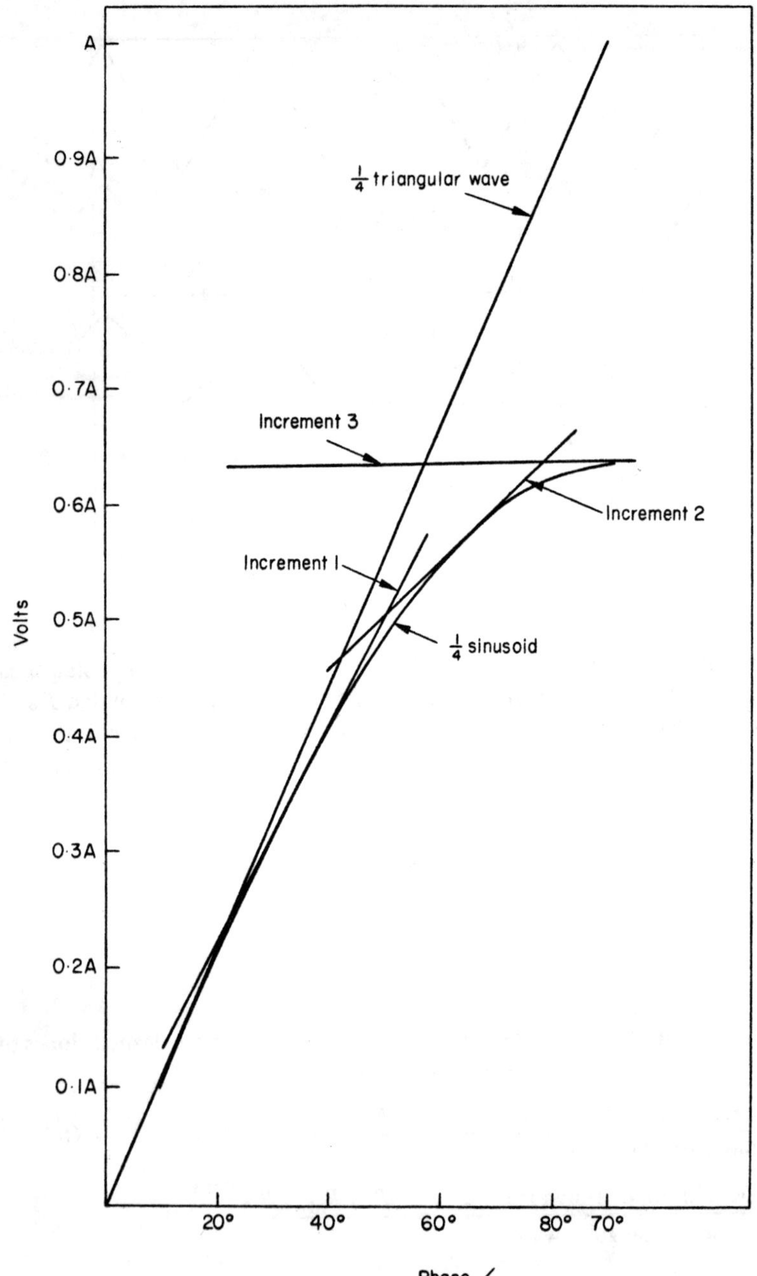

Fig. 6.11   Straight line approximated sinusoid

The values of the output voltage at which the changes of slope occur are determined by the equation

$$e_{on} = -E_{ref}\,\frac{R_{an}}{R_{bn}} \tag{6.6}$$

The above design equations can be used to calculate suitable resistor values for the feedback networks. Component values used in the circuit of figure 6.9 are based upon these equations; the resistors used are the approximate preferred values. If greater shaping accuracy is required resistor values should be trimmed and the sine wave should be approximated by a greater number of increments.

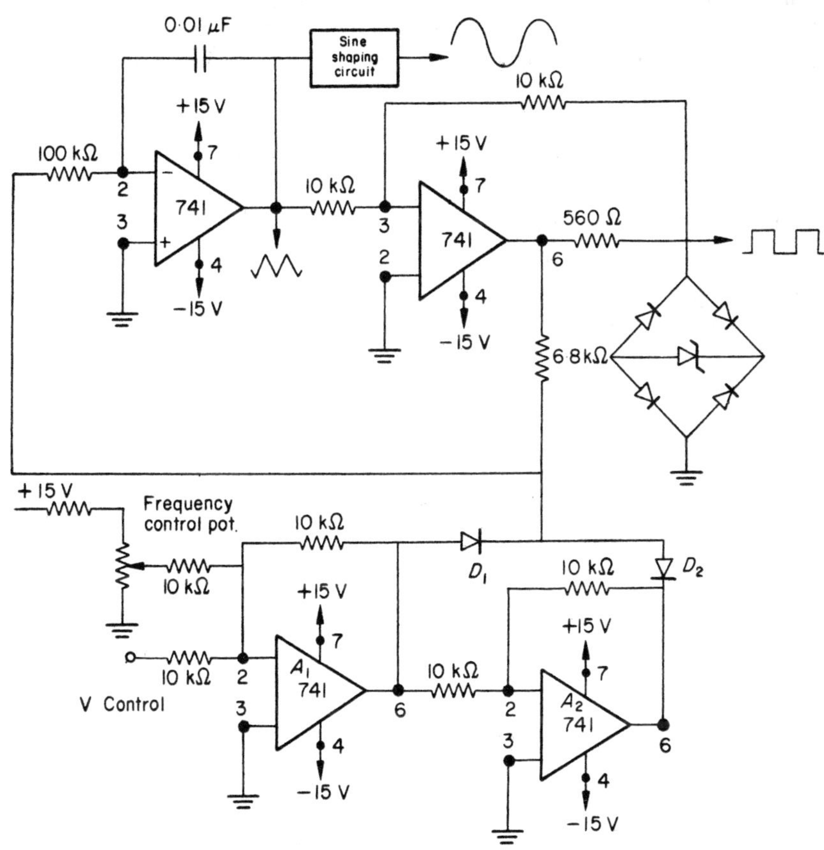

Fig. 6.12   Voltage controlled function generator

6.2.3 *Voltage Controlled Function Generator*

The versatility of a function generator is enhanced if the frequency of the signal which it produces can be controlled with a d.c. voltage or modulated with an a.c. voltage. A system which permits this form of control is illustrated in figure 6.12.

It will be remembered that in the basic function generator of figure 6.4 frequency is controlled by the amplitude of the square wave which is fed back to the integrator. In the circuit shown in figure 6.12 the positive and negative limits of the square wave applied to the integrator are clamped at the output levels of amplifiers A_2 and A_1 by diodes D_2 and D_1 respectively. Amplifier A_2 is connected as an inverting amplifier with its input signal supplied by the output signal of amplifier A_1. The output signals of both amplifiers and the frequency of oscillation of the function generator are thus controlled by the sum of the two input signals applied to amplifier A_1. One of these input signals is derived from a frequency control potentiometer the other is an external control voltage.

Some typical waveforms which may be obtained with the function generator system of figure 6.12 are shown in figures 6.13, 6.14 and 6.15. Figure 6.13 shows frequency modulation of the triangular wave at the integrator output produced in response to a sinusoidal control voltage applied to amplifier A_1. In figure 6.14 frequency is switched between two values by a square wave control signal, the signal displayed is that at the output of the sine shaping circuit. In figure 6.15 a ramp control signal is used to produce a frequency sweep.

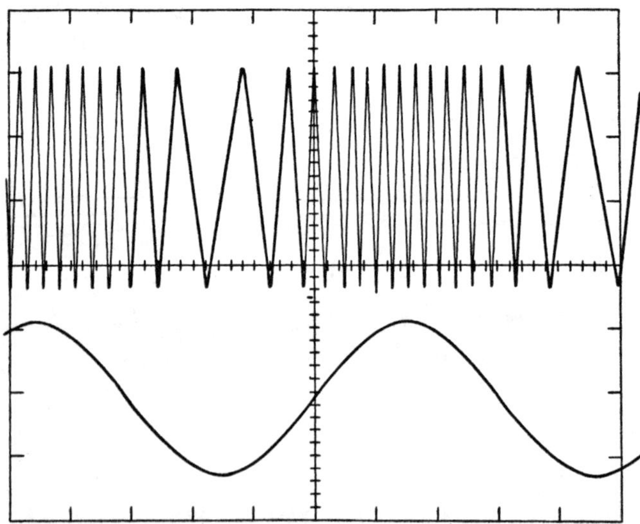

Fig. 6.13   Voltage controlled function generator frequency modulation; vertical 5V/division; horizontal 10ms/division

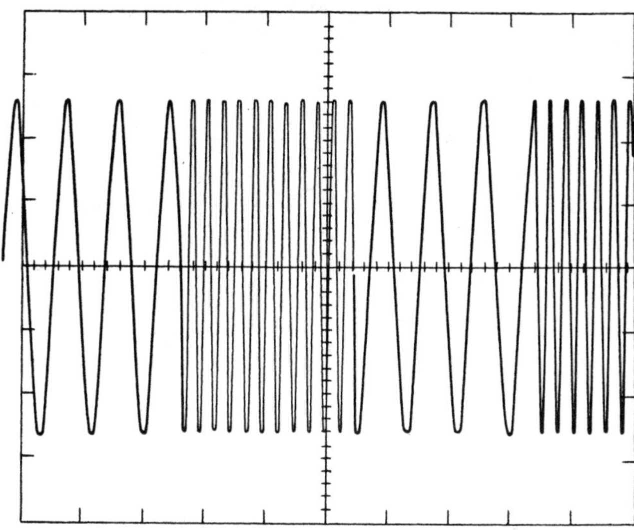

Fig. 6.14   Voltage controlled function generator switched frequency; time scale 10ms/division

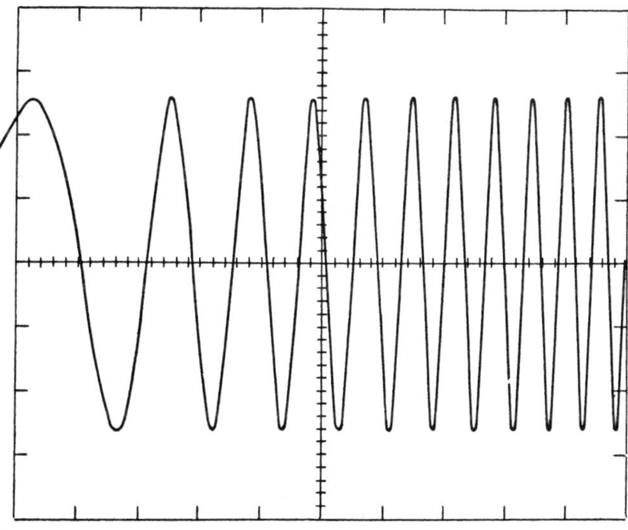

Fig. 6.15   Voltage controlled function generator swept frequency; time scale 10ms/division

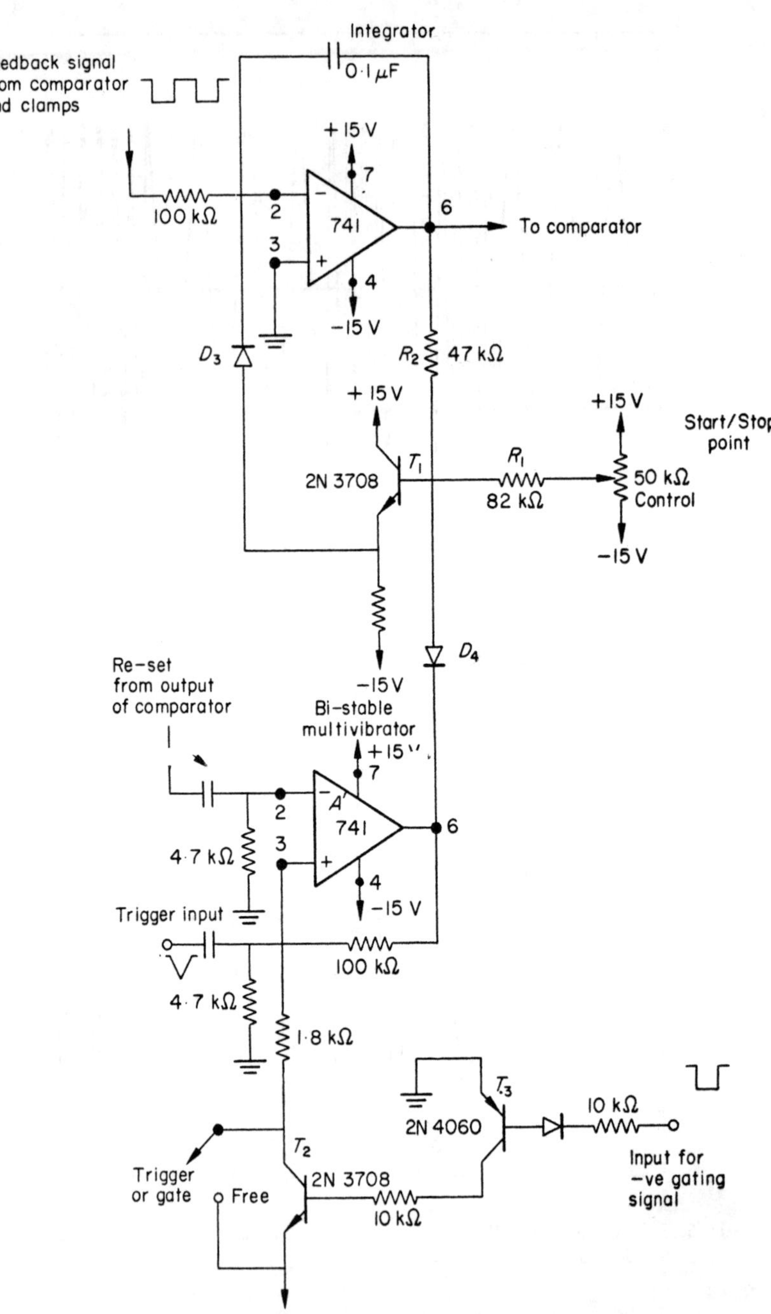

Fig. 6.16   Function generator modified to allow triggered or gated operation.
Remainder of circuit as in figure 6.12

6.2.4 *Triggered Function Generator*

Function generators of the type considered in this section, in which signal generation is performed by a closed loop consisting of an integrator and a comparator, can be readily modified to allow triggered or gated operation. The modification consists essentially of a circuit arrangement which interrupts the integrator action and holds the integrator output voltage at some predetermined level within its output range. Removal of the hold condition causes an immediate start of oscillations commencing at a point in the oscillation cycle corresponding to the hold level. Oscillations cease at this same point when the hold condition is re-applied. Since the closed loop feedback circuit which generates oscillations contains no tuned circuits there is no settling time required for the establishment of a stable oscillation amplitude and no ringing when oscillations cease.

Figure 6.16 shows a modification which can be made to the function generator system of figure 6.12 in order fo permit triggered or gated operation of the generator. The modification involves only the integrator section of the system. A hold condition is imposed on the integrator by the arrangement consisting of diode D_3 and transistor T_1, transistor T_1 being connected as an emitter follower. The integrator output is held at that level at which the sum of the integrator output voltage and a voltage derived from a start/stop point control, (scaled by resistors R_1 and R_2), causes D_3 to be forward biased. The hold condition is removed by a reverse biasing of diode D_3 caused by a forward biasing of diode D_4. The bias on diode D_4 is determined by the output voltage of amplifier A''. Amplifier A' acts as a bistable multivibrator; when its output level is positive diode D_4 is forward biased, diode D_3 reverse biased and the hold condition is removed.

In the triggered mode of operation of the circuit, a negative triggering pulse, applied to the non-phase inverting input terminal of amplifier A' causes a single cycle to be produced by the function generator. The action of the system is illustrated by the waveforms in figures 6.17 and 6.18. The traces in these figures show (from top to bottom): the externally applied trigger input, the bistable output, the voltage at the base of transistor T_1, the comparator output and the integrator output. The traces were obtained using two different settings of the start/stop point control towards opposite ends of its control range. The negative input trigger causes the bistable to switch to its negative input trigger causes the bistable to switch to its negative output state and oscillations commence starting at the hold level. At the point in the oscillation cycle at which the comparator switches to its negative output state a negative pulse derived from the comparator is applied to the phase inverting input terminal of the bistable.

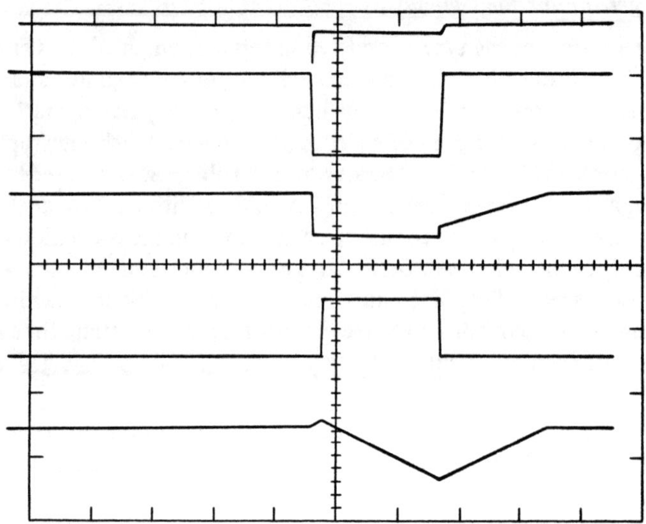

Fig. 6.17   Waveforms of triggered function generator; trigger 10V/division; other traces 10V/division; horizontal 1 ms/division

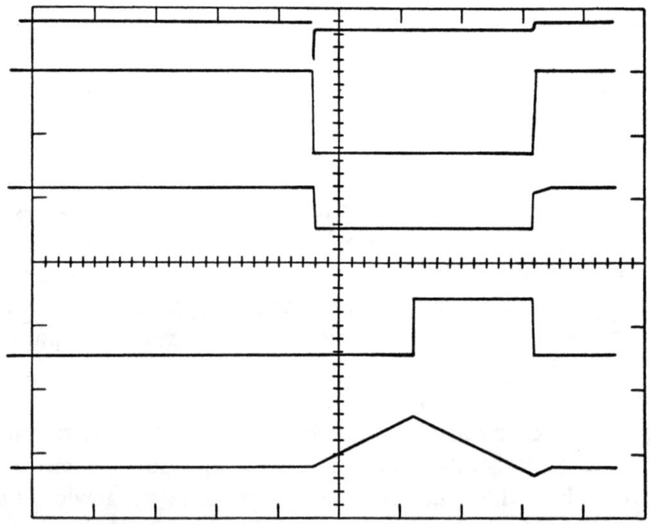

Fig. 6.18   Waveforms of triggered function generator (scales as before)

This pulse resets the bistable to its positive output state, the hold condition is re-imposed and oscillation ceases when the integrator output again reaches the hold level.

In the gated mode of operation of the system oscillations take place when transistor T_3 is switched on. The inclusion of the p-n-p transistor T_3 in the circuit allows an earth referred signal to be used for gating; the bistable is forced to remain in its negative output condition and the hold is removed for the duration of the gate signal. Oscillations start and end at a point in the oscillation cycle determined by the setting of the start/stop point control. The waveforms in figure 6.19 show a gated sinusoid at the output of the sine shaping circuit for two different settings of the start/stop point control.

In the free running mode of operation of the system the bistable is held permanently in its negative output state, the hold condition is not applied and the generator operates continuously.

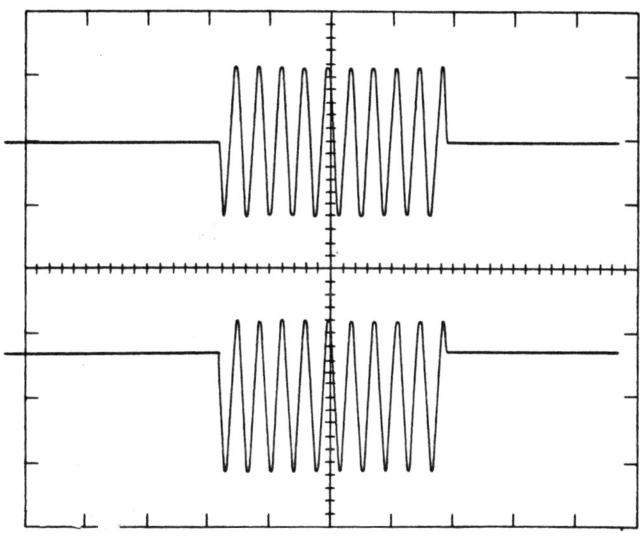

Fig. 6.19   Gated ouput with different settings of start/stop point; vertical scale 10V/division; horizontal 10ms/division

6.2.5 *Function Generator with Voltage Control of Amplitude*

A function generator which permits simultaneous voltage control of both amplitude and frequency is capable of producing a very wide range of wave shapes and is one of the most versatile signal sources available. The addition of a four-quadrant linear multiplier to the function generator system developed in the preceding sections gives the system a voltage control of amplitude capability. The multiplier is placed in the generator signal path immediately before the signal output point. Before applying the function generator signal to the multiplier it is advisable to bias multiplier offsets.

Some examples of the many different waveforms which can be generated by the complete function generator system are shown in figures 6.20 to 6.23. Figure 6.20 shows a sinusoidal signal with its amplitude switched between two defined values by a square wave control signal. Figure 6.21 shows amplitude modulation produced by a sinusoidal and a triangular control wave. Figure 6.22 shows a square wave control signal used to switch the phase of the output signal. In order to obtain this type of trace the output amplitude control potentiometer, *P*, is adjusted so that there is zero output when no control signal is applied. Figure 6.23 shows a simultaneous amplitude and frequency sweep with the same ramp control signal used to vary both the amplitude and frequency of the generator signal.

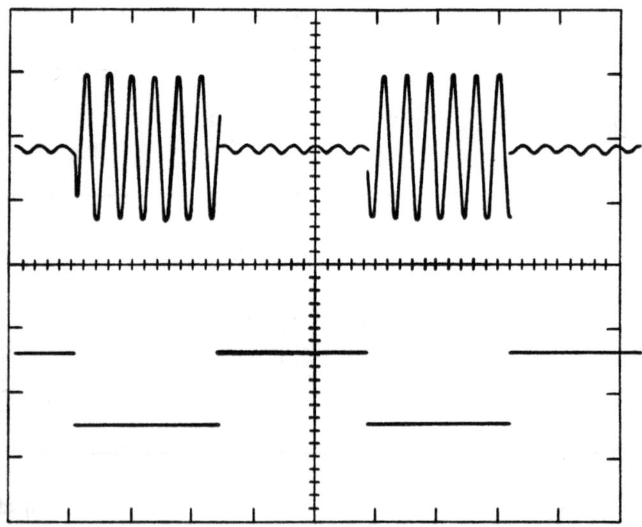

Fig. 6.20   Function generator square voltage used to control amplitude; vertical 5V/division; horizontal 10ms/division

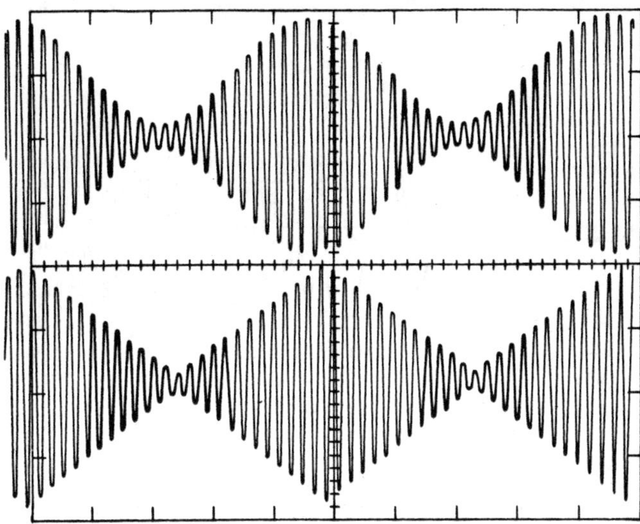

Fig. 6.21 Function generator amplitude modulation produced by sinusoidal and triangular wave; horizontal scale 20ms/division

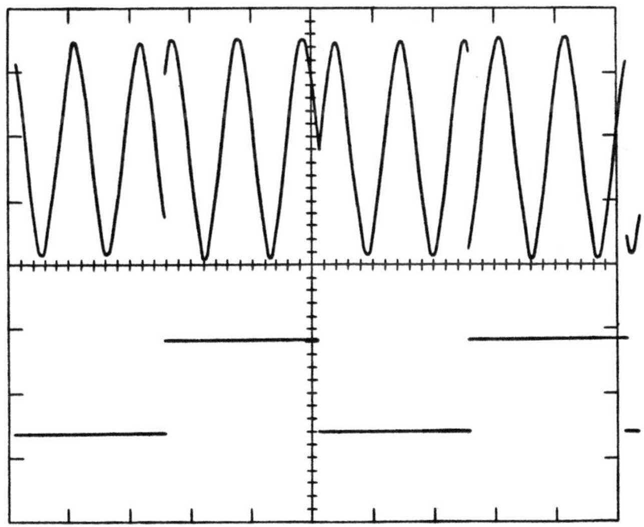

Fig. 6.22 Function generator square wave controlled signal used to switch phase; vertical 5V/division squarewave, 2V/division sinusoid; horizontal 5ms/division

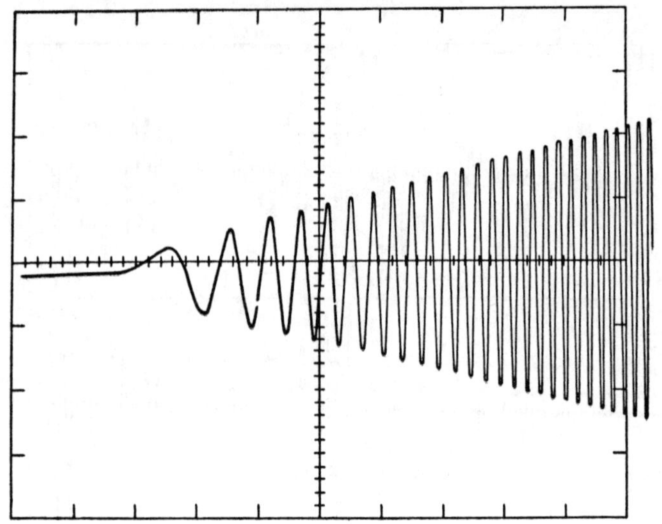

Fig. 6.23   Function generator swept amplitude and frequency; vertical 1V/division horizontal 20ms/division

Extremely precise waveform generator systems can be developed from the circuits that have been dealt with in the preceding sections, performance errors can be made very small by appropriate choice of amplifier types and the use of precision components. Upper frequency limitations exist because of amplifier slewing rates, lower frequency limitations are set by integrator drift; sawtooth linearity error is also determined by integrator performance. As an alternative to the synthesis of a function generator system from individual function units there are now several monolithic waveform generators available. These devices allow considerable reduction in external circuit complexity and should be considered for use in waveform generator applications not requiring the ultimate in precision performance.

Exercises 6

6.1   The basic function generator system shown in figure 6.4 uses the following component values: $R_1 = 10$ kΩ, $R_2 = 20$ kΩ, integrator capacitor $= 0.02$ μF, integrator resistor $= 100$ kΩ. The output limits of the comparator are determined by the amplifier saturation limits of ($\pm$) 10 volts.
What will be the maximum frequency of oscillations produced by the circuit? Sketch the waveforms to be expected. If the amplifier has the input offset voltage and bias current given in Exercises 2 at what frequency will integrator drift cause 5 per cent error in waveforms?

Appendix

Operational Amplifier Performance Errors [1,2]

1. *Gain Bandwidth Errors*

In negative feedback circuits the assumed infinite open loop gain of the ideal operational amplifier makes closed loop performance characteristics dependent only upon the magnitude of the external components used to determine the feedback fraction β. The effect of finite open loop gain is to introduce a so called gain error factor such that

$$\begin{matrix} \text{Actual closed loop} \\ \text{transfer function} \end{matrix} = \begin{matrix} \text{Ideal closed loop} \\ \text{transfer function} \end{matrix} \left[\frac{1}{1 + \dfrac{1}{\beta A_{OL}}} \right]$$

$$\text{The gain error factor} = \frac{1}{1 + \dfrac{1}{\beta A_{OL}}}$$

If the loop gain $-\beta A_{OL}$ is real, negative and much greater than unity, performance errors due to finite A_{OL} are usually less than those due to the tolerance in external component magnitudes.

Loop gain is frequency dependent; this dependence may be represented graphically by drawing the Bode magnitude plots for A_{OL_ω} and $1/\beta_\omega$

$$|\beta A_{OL(\omega)}| \text{ in dB} = |A_{OL(\omega)}| \text{ in dB} - |1/\beta| \text{ in dB}$$

and the magnitude of the loop gain is represented by the difference in the Bode plots.

Both the magnitude of the loop gain and its phase are equally important in determining the gain error factor. If the phase shift in the loop gain does not exceed $90°$ the magnitude of the gain error factor is always less than unity, but phase shifts greater than $90°$ cause the gain error factor to take on values greater than unity at the frequency at which the magnitude of the loop gain is unity. An estimate of the phase shift in the loop gain at the frequency at which the magnitude of the loop gain is unity can be made from the rate of closure of the Bode plots. If the rate of closure does not exceed 20 dB per decade the phase shift does not exceed $90°$ and the gain error factor is less than unity at all frequencies. A rate of closure of 40 dB per decade implies a phase shift of $180°$ which causes closed loop instability.

As an example we consider an operational amplifier, with a first order frequency response, connected in the follower configuration. The circuit and its Bode plots are illustrated in figure A1. If $\beta A_{OL(o)} \gg 1$ the $|A_{OL(\omega)}|$ and $|1/\beta|$ plots intersect at a frequency $\omega_1 = \beta A_{OL(o)} \omega_c$ at which frequency the magnitude of the loop gain is unity and the magnitude of the gain error factor is $1/\sqrt{2}$. ω_1 represents the closed loop bandwidth. Note that the closed loop gain bandwidth product remains constant at a value equal to the open loop gain bandwidth product.

As a second example of the use of Bode plots in evaluating closed loop frequency response characteristics we consider the simple differentiator circuit shown in figure A2, (see section 2.4). The ideal differentiator transfer function is

$$A_{(p)} = -pCR$$

but marked errors occur at frequencies for which the magnitude of the loop gain is unity.

$$\beta = 1/(1+pCR), \quad 1/\beta = 1+pCR,$$

Bode magnitude plots are illustrated. Note that when

$$|A_{OL(\omega)}| \quad \text{and} \quad |1/\beta_{(\omega)}| \quad \text{close at a rate}$$

of 40 dB/decade the phase shift in $\beta A_{OL(\omega)}$ exceeds $90°$ and the gain error factor takes on values greater than unity. Phase shift, at the crossover frequency, can be estimated from the Bode phase approximations and can be used to find a value of the gain error factor at this frequency. Alternatively, we may write the differentiator transfer function as

$$A_{(p)} = -pCR \left[\frac{\dfrac{1}{}}{1 + \dfrac{1}{\beta A_{OL(p)}}} \right]$$

substitution for $\beta_{(p)}$ and $A_{OL(p)}$ and rearrangement gives

$$A_{(p)} = -\frac{p\, A_{OL(o)}\, \omega_c}{p^2 + p\left[\omega_c + \dfrac{1}{CR}\right] + \dfrac{\omega_c}{CR}\left[A_{OL(o)} + 1\right]}$$

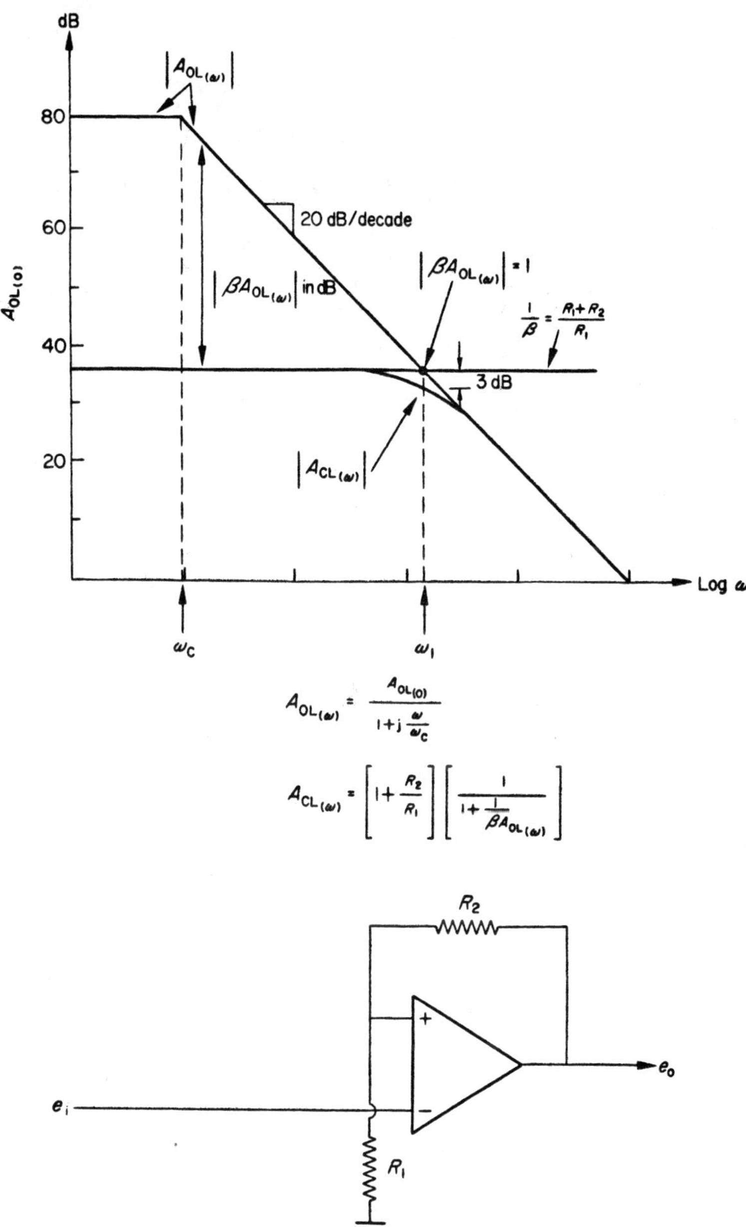

Fig. A.1   Bode plots for follower

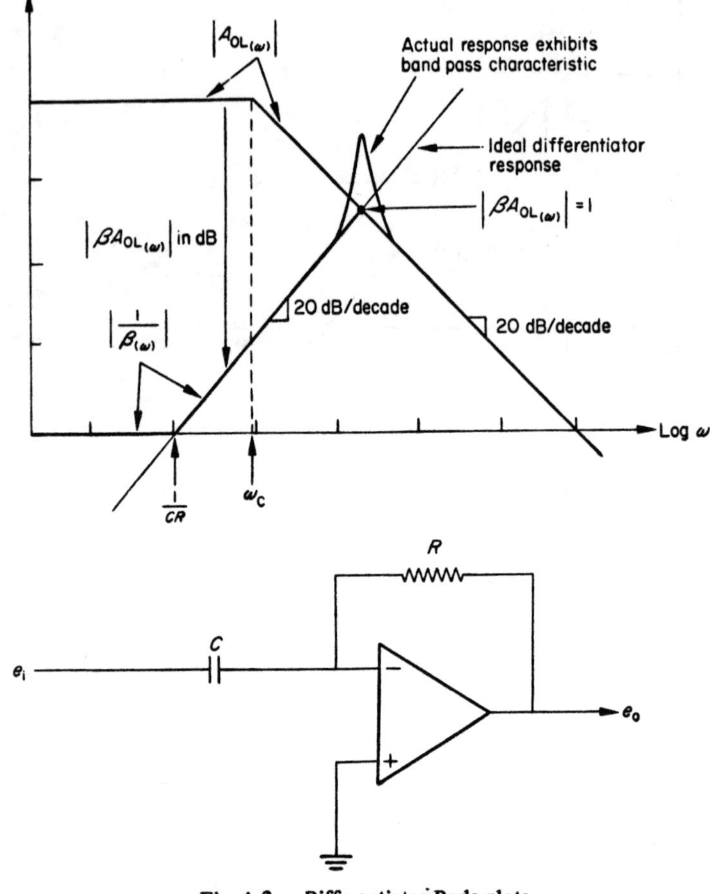

Fig. A.2   Differentiator Bode plots

The equation represents a second order band pass function, with damping factor

$$\zeta = \tfrac{1}{2}\left(\omega_c + \frac{1}{CR}\right)\sqrt{\frac{CR}{\omega_c\,(A_{OL(o)} + 1)}}$$

and natural frequency

$$\omega_0 = \sqrt{\frac{\omega_c}{CR}\,(A_{OL(o)} + 1)}$$

If the damping factor is small ($\zeta < 1/\sqrt{2}$) the differentiator transient response is characterised by a marked 'ringing'.

The differentiator can be frequency compensated by connecting a resistor R_1 in series with input capacitor C. The modified Bode plot for $1/\beta$ is shown in figure A3. If the value of R_1 is suitably chosen

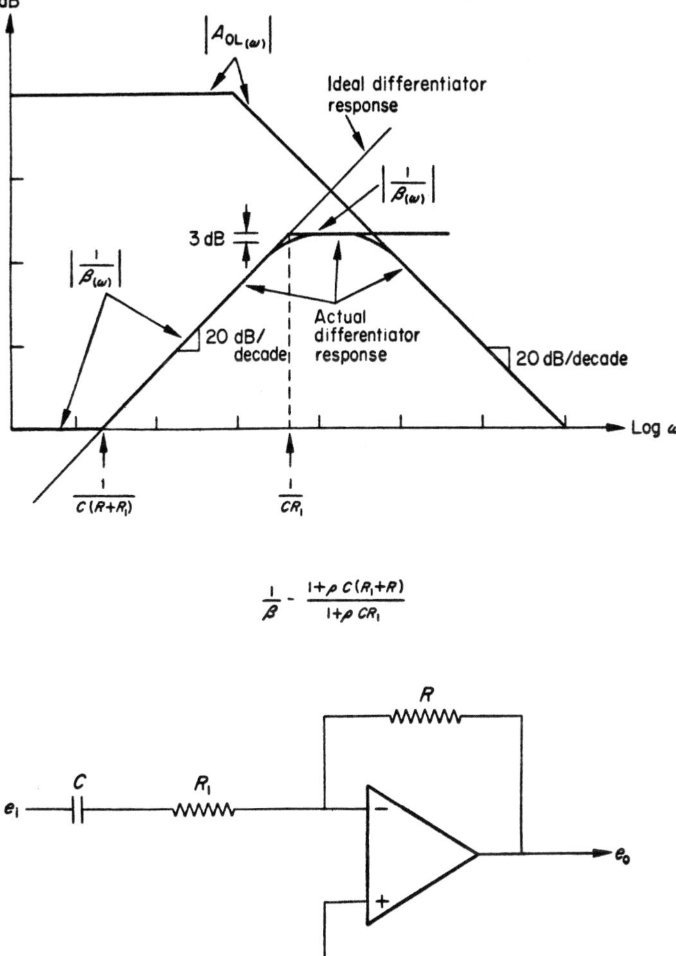

Fig. A.3   Bode plots for differentiator with frequency compensation

$$\left|\frac{1}{\beta_{(\omega)}}\right| \text{ and } \left|A_{\mathrm{OL}(\omega)}\right|$$

close at a rate of 20 dB per decade and the circuit reponse is damped. Note that the differentiator sinusoidal response is in error at frequencies above $1/C R_1$, it is 3 dB in error at the frequency $1/C R_1$.

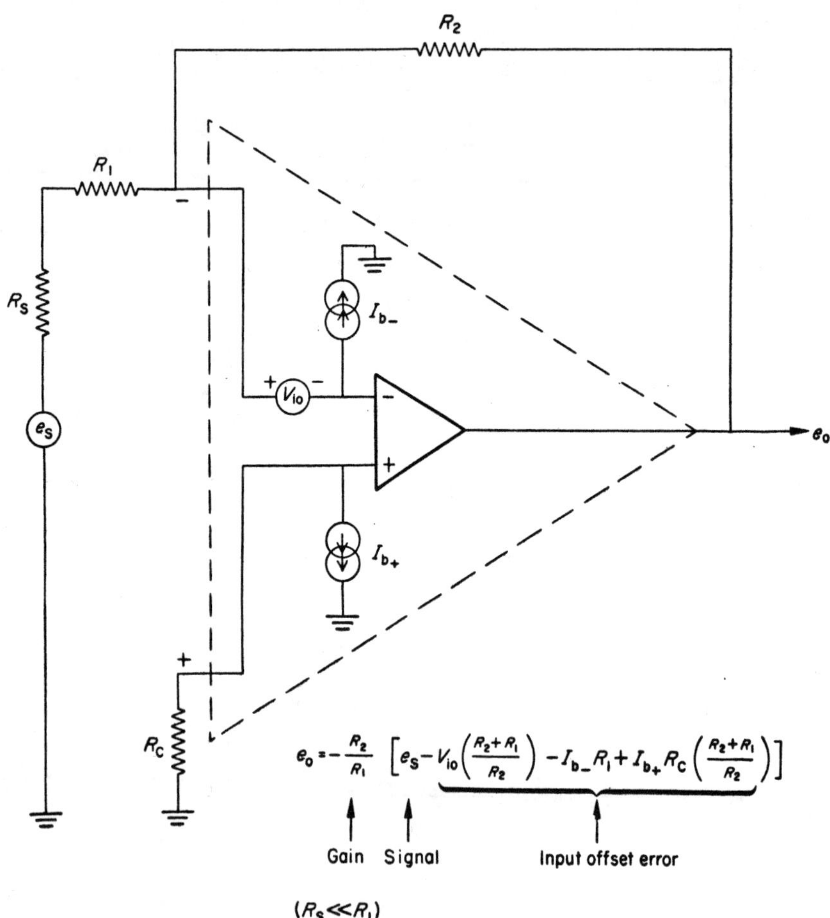

$$e_0 = -\frac{R_2}{R_1}\left[\,\underbrace{e_S}-\underbrace{V_{io}\left(\frac{R_2+R_1}{R_2}\right) - I_{b-}R_1 + I_{b+}R_C\left(\frac{R_2+R_1}{R_2}\right)}\,\right]$$

Gain   Signal   Input offset error

$$(R_S \ll R_1)$$

Fig. A.4   Input offset error for inverting configuration

2. *Offset and Drift Errors*

Offset errors arise in operational amplifier applications as a result of amplifier bias current flowing through input source resistances and as a result of the amplifier's own input offset voltage. The errors can be analysed by representing the amplifier offsets in terms of error generators at the input of an ideal amplifier. The results of such analysis are given, for the inverter in figure A4, for the follower in figure A5 and for the current to voltage converter in figure A6.

Amplifier drift causes a change in offset error, the change in offset error represents the drift error. Temperature drift is usually the most significant error.

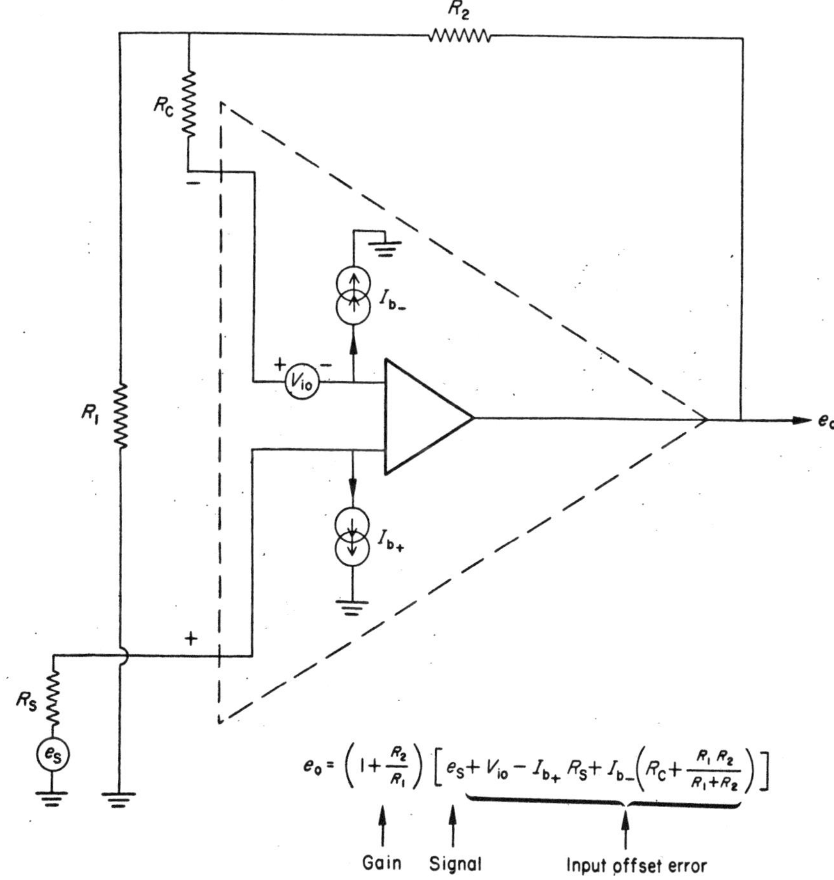

$$e_0 = \left(1+\frac{R_2}{R_1}\right)\left[e_S + V_{io} - I_{b_+}R_S + I_{b_-}\left(R_C + \frac{R_1 R_2}{R_1+R_2}\right)\right]$$

Fig. A.5   Input offset error for non-inverting configuration

The temperature drift error for a temperature change δT can be found by substituting

$$V_{io} = \frac{\Delta V_{io}}{\Delta T}\,\delta T, \quad I_b = \frac{\Delta I_b}{\Delta T}\,\delta T$$

$$I_{io} = I_{b+} - I_{b-} = \frac{\Delta I_{io}}{\Delta T}\,\delta T$$

in the appropriate input offset error expressions.

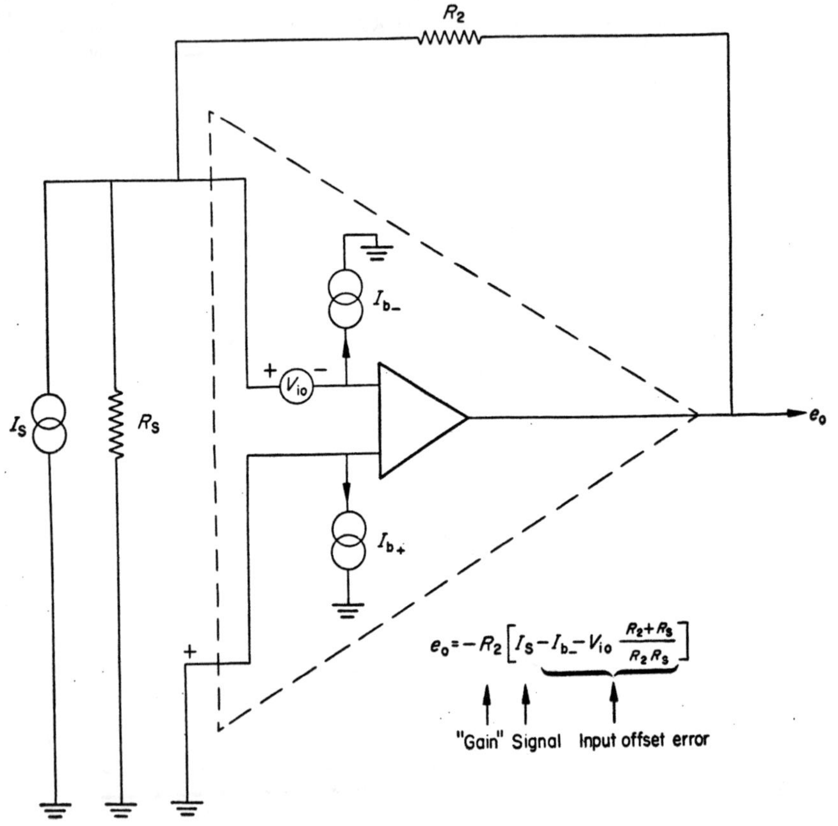

Fig. A.6   Input offset error for current to voltage converter

3. *Common Mode Error*

In circuit configurations in which a common mode input voltage is applied to an amplifier, a non-infinite amplifier CMRR gives rise to performance errors. Errors can be calculated in terms of an equivalent input error generator of magnitude

$$e_{\Sigma cm} = \frac{e_{cm}}{CMRR}$$

The representation is illustrated for a simple follower configuration in figure A7.

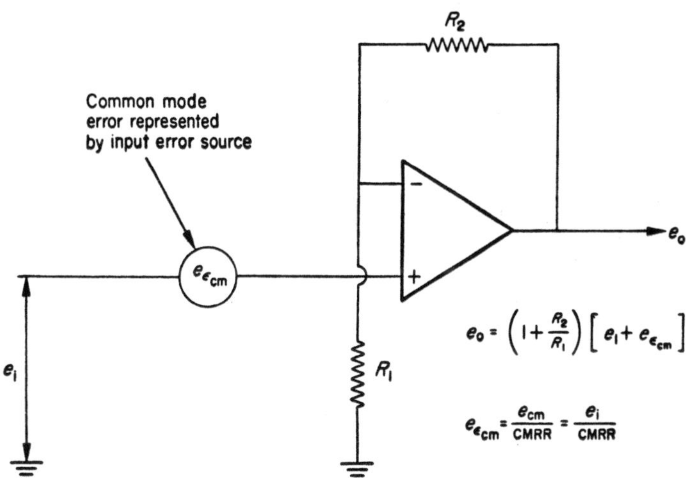

Fig. A.7   Representation of common mode error

References

1. G.B. Clayton, *Operational Amplifiers*. Butterworths (1971).

2. Tobey, Graeme, Huelsman (Barr Brown), *Operational Amplifiers Design and Applications*. McGraw Hill (1971).

Answers to Exercises

Exercises 2

2.1 (a) 10^5 rad/s
      (b) 40 dB, 1%
      (c) 0.7 V
      (d) 0.02 V
      (e) 0.02 V
      (f) 9.9 kΩ, 0.25 V, 0.011 V, 0.011 V

2.3 (a) 7.94×10^{-3}
      (b) 7.94×10^4 rad/s
      (c) 28 dB, 40 dB
      (d) 1.3%
      (e) 0.5 V
      (f) 0.018 V
      (g) 0.72 mV, 0.18 mV
      (h) 3.97 kΩ, 0.28 V, 0.013 V, 0.52 mV, 0.13 mV

2.4 0.01%

2.5 (a) 51 μA
      (b) 1 μA

2.6 -3 V

2.7 (a) 1 V/s
      (b) 2×10^{-2} V/s

2.8 5%

Exercises 3

3.1 $R_{a1} = 50$ kΩ $R_{b1} = 250$ kΩ
      $R_{a2} = 100$ kΩ $R_{b2} = 33$ kΩ
      $R_{a3} = 50$ kΩ $R_{b3} = 500$ kΩ
      $R_{a4} = 100$ kΩ $R_{b4} = 143$ kΩ

3.2 $R_2 = 1.57\ \text{k}\Omega$, R_1 say 1 MΩ

3.3 $R_4 = 15.7\ \text{k}\Omega$ $e_2 = 4$ V

3.4 $R_7 = 9.66$ K $e_3 = 2.83$ V

Exercises 5

5.1 ± 5 V

5.2 $C = 0.01\ \mu\text{F}$, $R_3 = 22\ \text{k}\Omega$ $R_4 = 68\ \text{k}\Omega$

5.3   0.001 seconds per volt

5.4 10^3 volt per second

5.6 1.1×10^2 Hz per volt, $e_1 \leqslant 1.04$ volts

Exercises 6

6.1   250 Hz   25 Hz

Index